John Gribbin gained a PhD from the Institute of Astronomy in Cambridge (then under the leadership of Fred Hoyle), before working as a science journalist for *Nature* and later *New Scientist*. **Mary Gribbin** is a teacher with a special gift for communicating difficult concepts, and she is a previous winner of the TES Junior Information Book Award. They have co-written several titles for adults and children. John has written many bestselling popular science books, including *Erwin Schrödinger and the Quantum Revolution*, *In Search of the Multiverse* and *The Universe: A Biography*. John and Mary are both Visiting Fellows at the University of Sussex.

'Elegant and meticulously researched . . . this is a most enjoyable book.' *The Observatory*

Praise for the authors:

'Mary and John Gribbin write with great clarity.' *Guardian*

'Precise yet mysterious . . . as beautiful as a poem and as exciting as a novel.' *Sunday Times*

'A gripping account of the history of quantum mechanics and a clear description of its significance – and weirdness. Absolutely fascinating.' *Isaac Asimov*

'Immensely readable.' *The Times*

OUT OF THE SHADOW OF A GIANT

HOW NEWTON STOOD ON THE SHOULDERS OF HOOKE AND HALLEY

JOHN GRIBBIN
AND
MARY GRIBBIN

WILLIAM
COLLINS

William Collins
An imprint of HarperCollins*Publishers*
1 London Bridge Street
London SE1 9GF

WilliamCollinsBooks.com

This William Collins paperback edition published in 2018

23 22 21 20 19 18
10 9 8 7 6 5 4 3 2 1

First published in Great Britain by William Collins in 2017

A catalogue record for this book
is available from the British Library.

ISBN 978-0-00-822061-7

Printed and bound in Great Britain by
CPI Group (UK) Ltd, Croydon, CR0 4YY

MIX
Paper from
responsible sources

FSC
www.fsc.org   **FSC` C007454**

This book is produced from independently certified FSC™ paper
to ensure responsible forest management.

For more information visit: www.harpercollins.co.uk/green

PREFACE

The seed from which the idea for this book grew was planted during a conversation with Lisa Jardine at the Royal Society, following a talk by one of us (JG). We got to speculating about how science in Britain might have developed if Isaac Newton had never lived. Our conclusion, such as it was, was that although Newton had inspired a great advance, and fully justified his status as the scientific giant of his day, there were only slightly lesser men who would have been well able to set British science off on the road it followed after Newton, although the journey down that road might have taken a little longer. Two men, in particular, stand out as thinkers who made major contributions, not just to scientific discovery but also to the development of the scientific method itself, who lived and worked in the shadow of Newton. They have by no means been forgotten, but even many of the people who still know the names of Robert Hooke and Edmond Halley have little knowledge of the remarkable breadth and depth of their work. Hooke is remembered for a rather mundane 'law' describing the behaviour of a stretched spring; Halley for the comet that bears his name, but which he did not discover. Their other achievements, however, are so important that between them they arguably add up to the scientific equivalent of another Newton. So rather belatedly (and, alas, too late for Lisa Jardine to see it) we have decided to attempt to bring them out from the shadow of Newton, and present the men and their achievements in all their glory.

ACKNOWLEDGEMENTS

Thanks to the University of Sussex for providing us with a base from which to work, the Royal Society, Royal Astronomical Society, British Library, British Museum, Public Record Office, Science Museum Library Wroughton, King's College, Cambridge, and the university libraries in Oxford and Cambridge for access to their archives, and to the Alfred C. Munger Foundation for financial support.

CONTENTS

INTRODUCTION

OUT OF THE
SHADOWS

Isaac Newton famously commented that if he had seen further than other people it was 'by standing on the shoulders of giants'. But even within his own lifetime, and increasingly since then, Newton was widely acknowledged as the greatest of all scientific giants, to such an extent that the remarkable achievements of his colleagues and contemporaries are often overlooked. Two of the pioneering scientists who lived and worked in the shadow of Newton would each have been regarded as giants in their own right if he had not been around, and it is our intention to bring them out of Newton's shadow to put their achievements in perspective. They are (in chronological order) Robert Hooke (1635–1703), who was slightly older than Newton (1642–1727), and Edmond Halley (1656–1742), who outlived Newton. Their overlapping lives neatly embrace the hundred years or so during which science as we know it became established in Britain.

But what of Newton? He was, to say the least, economical with the truth, and attempted to write Robert Hooke out of history, having 'borrowed' many of Hooke's best ideas. It has been established, for example, that the famous story of the falling apple seen during the plague year of 1665 is a myth, invented by Newton to bolster his (false) claim that he had the idea of a universal theory of gravity before Hooke. In fact, Hooke described such an idea, and the rule that every object (such as a planet) moves in a straight line unless acted upon by some outside force (ironically, now known as Newton's First Law), in the mid-1660s, when Newton was an unknown and very junior member of Cambridge University (he only graduated in 1665). Until Hooke mentioned these ideas to Newton several years later, Newton subscribed – as surviving documents show – to the idea that planetary motion was caused by whirlpools in some kind of fluid filling the Universe. Newton also lifted much of Hooke's work on light and colours, and Newton published (significantly, immediately after Hooke's death) as 'his' own theory of heat, an idea that has been described by historian Clara de Milt as 'very, very' like Hooke's earlier work. In one respect, though, Newton was a better scientist than Hooke: he was a brilliant mathematician. And he outlived Hooke, so he had the last word – until now.

We have not attempted to provide complete biographies of our subjects, who have been well served in that regard by Lisa Jardine (Hooke) and Alan Cook (Halley); our focus is on their scientific achievements, and how these were fundamental to the development of science in England. But there is, we hope, enough biographical background here to give some insight into the kind of people they were, and how they were both, in their different ways, products of the society they lived in.

Hooke has been described as 'England's Leonardo', a polymath whose achievements extended far beyond the realm of science. He came from a poor but respectable background, the son of a curate on the Isle of Wight. With the aid of a modest inheritance, he was able to attend Westminster School, and go on to Oxford as a choral scholar without having to pay fees. This was in 1653, during the Parliamentary Interregnum, and since music was banned in church services at the time he got the scholarship without having to sing for it. He eked out a living by acting as a paid servant for one of the wealthier undergraduates, normal practice at that time. He then moved on to become assistant to Robert Boyle, the 'father of chemistry'. It is now widely accepted that it was Hooke who discovered what is now known as 'Boyle's Law' of gases; Hooke was the only one of Boyle's various paid assistants to be credited by name in his writings. Through Boyle, Hooke became a member of the inner circle of British scientists of the day, and became the first Curator of Experiments at the Royal Society. He was the man who made the Society a success, but (perhaps because of his humble background) he was always touchy about priority and famously got involved in rows with Newton and the Dutch scientist Christiaan Huygens. Hooke pioneered the use of the microscope, wrote the first popular science book (praised by Samuel Pepys), made astronomical observations, and kept the Royal running. One of his great friends was Christopher Wren, and after the Fire of London the two of them worked together on the rebuilding of the City – many 'Wren' churches are now thought to be Hooke's work. Hooke was the best experimental scientist of his time, the leading microscopist of the seventeenth century, an astronomer of the first rank, and developed an understanding of earthquakes, fossils and the history of the Earth that would not be surpassed for a century.

Edmond Halley is remembered today for the comet that bears his name, but, like Hooke, he had several strings to his bow. Halley came from a relatively prosperous background, did well at school, and went up to Oxford in 1673. There, he was able to indulge his passion for astronomy by taking with him equipment including a telescope and sextant that would have been the envy of a contemporary professional astronomer. His work impressed John Flamsteed, the first Astronomer Royal, and with his (and other) help Halley was able to wangle a trip to the island of St Helena to carry out a survey of the southern skies. His father provided an allowance of £300 a year (three times Flamsteed's salary!), the degree was abandoned, and at the age of twenty Halley went off on his adventure. The survey was a great success, with Halley's *Catalogue of the Southern Stars* establishing his reputation. The King 'recommended' that Halley be given a degree, which was awarded three days after he was elected as a Fellow of the Royal Society. He was not, however, in a hurry to build on this success. Comfortably supported by his father, he led the life of a gentleman in Restoration England, including the Grand Tour of Europe, before settling down, getting married and publishing astronomical observations from his home in Islington. The death of his father (in suspicious circumstances) in 1684 brought a change of priorities, and Halley became more involved with the work of the Royal Society. It was around this time that Halley, Hooke and Wren, puzzling over the nature of planetary orbits, asked Newton if he could explain why they seemed to be governed by an inverse square law. This led to the publication, overseen and funded by Halley, of Newton's great work, the *Principia*. Alongside all this, Halley carried out a survey of the Thames estuary and invented a practical diving bell. He made the first scientific estimate of the size of atoms, calculated how to work out the distance to the Sun from a transit of Venus, and set out on an official voyage of exploration to the southern seas – a predecessor of the famous *Beagle* voyage. But, unlike Darwin, he was not a mere passenger; he was given a King's ship and made Master and Commander (in modern language, a Royal Navy Captain) to run it, the only 'landsman' ever to hold such a post. This led to secret work as a spy (the details are lost) in the Adriatic, and then his appointment as a Professor at Oxford University. His life became less exciting, although it was recorded that he 'talks, swears, and drinks brandy like a sea captain'. But there remained many scientific contributions, including the prediction of the return of 'his' comet and Halley's appointment as the second Astronomer Royal, in 1720.

So far, so good. This outline is essentially the story we expected to tell. But as we delved into the historical material, we found that the importance of Hooke and Halley is even greater than we had anticipated, while Newton turned out to have feet of clay. Newton got some of his best ideas – including 'Newton's First

Law' of motion, and the idea of gravity as a universal attractive force – from Hooke, and shamelessly took credit for them. He is known to have lied about his priority more than once, and to have deliberately tried to write Hooke out of the story. But he was only in a position to do so thanks to Halley. Preferring the quiet life as a reclusive Cambridge academic to the rough and tumble of scientific debate in Restoration England,* Newton would have remained an obscure minor figure, remembered in the footnotes of science only for his (incorrect) theory of light, if Halley had not first prodded him into writing his masterpiece, the *Principia*, and then paid for its publication out of his own pocket. Without Hooke and Halley, we might never have heard of Newton. Without Newton, we would have heard a lot more about Hooke, in particular, and Halley. The legend that grew up about Newton was largely Newton's invention, and became regarded as fact. For some three hundred years Newton has been venerated in the spirit of the famous line from the movie *The Man Who Shot Liberty Valence* – 'When the legend becomes fact, print the legend.' But it is our intention to print the facts.

We tell the story from the perspective of the intertwined lives of Hooke and Halley, from 1635–1742, starting with the birth of Hooke and taking his story forward, then picking up Halley's story and carrying both forward. The greatest overlap concerns the time when Hooke and Halley were involved in stimulating Newton's greatest work, which we describe both from Hooke's perspective (Chapter Seven) and from Halley's perspective (Chapter Eight); this inevitably involves some repetition, but by dealing with this from the different perspectives we hope to make their intertwined relationships clear. Along the way, we also describe their interactions with other scientists, not just Newton. And we will leave them, we hope, basking in the sunlight of the recognition they deserve.

The key development in seventeenth-century science, certainly

---

* Newton had good reason to be reclusive, given the times he lived in; he was almost certainly homosexual, and definitely a religious heretic.

in Britain and arguably in the world, was the establishment of the Royal Society in the 1660s. It was through the Royal Society that Hooke, Halley and Newton met and interacted with each other. Before then, there had been individual scientific pioneers, notably the philosopher Francis Bacon and the experimenters William Gilbert and Galileo Galilei. But the Royal provided a forum for those of a scientific bent to meet, discuss ideas and experiment, as well as being a kind of clearing house for scientific information gathered through a network of correspondents. It was Robert Hooke, more than anyone else, who made the society a success in its early days, when without his enthusiasm, hard work and skill it might have foundered. But to put this in context, we need to begin before Newton, Halley or the Royal Society were even conceived, as Hooke's early life brings out the changes in British society around the time of the Civil War and paved the way for a scientific revolution.

John Gribbin
Mary Gribbin
November 2016

A NOTE ON DATES

Until 1752, the English used the old Julian calendar, which was ten days behind the Gregorian calendar (our modern calendar) used across mainland Europe, which had been introduced because the calendar dates on the old calendar had gradually slipped out of sequence with the seasons. Hooke and Halley therefore used Julian ('Old Style' or OS) dates, and we have kept these except where we have indicated 'New Style' (NS). At the time, the New Year officially began on 25 March, the start of the tax year, but most people, as today, regarded 1 January as the start of the year. Astronomers, in particular, dated the new year from 1 January, and as both Robert Hooke and Edmond Halley were astronomers, that is good enough for us. In some sources, dates between 1 January and 24 March are written with both numbers, as for example 1650/1, but we give all our dates assuming the year began on 1 January.

CHAPTER ONE

FROM FRESHWATER
TO OXFORD

According to his own account, Robert Hooke was born at exactly noon on 18 July 1635, at Freshwater on the Isle of Wight. What we know of his early life comes from two sources. John Aubrey's *Brief Lives* is always entertaining, although not always accurate, but Aubrey was a friend of Hooke and had many conversations with him. Another friend, the naturalist Richard Waller, knew Hooke in later life, and was responsible for publishing the *Posthumous Works of Robert Hooke* in 1705, putting into print some of Hooke's previously unpublished lectures. Waller's introduction to that book drew, he tells us, on an autobiographical memoir that Hooke started to write but never finished, and which is now lost. The story pieced together from these two sources can be fleshed out, however, with other information about events on the Isle of Wight in particular, and across Britain in general, at the time Hooke was growing up. It was certainly an interesting time to be alive, taking in civil war and the execution of a king before the boy was fourteen.

Hooke's father, John, was the curate of All Saints Church,

where the rector was Cardell Goodman, a staunch Royalist and former member of Westminster School and Christ Church, Oxford – connections that would in due course become important to Robert. His mother, Cecellie, was the second wife of John Hooke, presumably a good deal younger than him, and Robert was by some way the youngest of four children. He had two sisters, the younger of whom was seven years older than him, and a brother, John junior, born in 1630.

Robert was a sickly baby who was christened the day after his birth, probably because he was not expected to live, but he survived to become a sickly child. He was too delicate to be sent away to school in Newport like his brother but was educated at home by his father. Although plagued by recurrent headaches and other ailments, this left him plenty of time to wander the south-west corner of the island, gradually becoming stronger, and to follow his own interests, which leaned towards practical activities such as making working models. These demonstrated a rare skill at an early age. He built a model ship, described by Waller as 'about Yard long, fitly shaping it, adding its Rigging of Ropes, Pullies, Masts, &c. with a contrivance to make it fire off some small Guns, as it was sailing cross a Haven of pretty breadth' (probably Yarmouth). When he saw a brass clock that had been taken to pieces for repairs, he copied the components in wood and put them together to make a clock, which worked tolerably well. The downside of all the hours he spent over a workbench was that by the time he was sixteen Robert had, he told Waller, developed a pronounced stoop, sometimes referred to by his biographers as a hunchback. As these examples show, Robert Hooke was a precocious 'mechanic' of rare skill. But his skills extended beyond the practical. When the artist John Hoskins, a painter at the court of Charles I, visited the island, Robert watched him at work, then made his own paints from materials to hand, such as coal, chalk and an iron ore known as ruddle, and used them to copy paintings hanging in the house, to such good effect that Hoskins suggested he could have a career as an artist.

The other formative influence on the young Hooke was the world around him. That part of the Isle of Wight offers spectacular scenery, chalk cliffs, and the dramatic sight of the Needles, a series of chalk spires rising from the sea at the end of a chalky spine running across the island. Many of the island strata are rich in fossils. Even as a child, Robert was intrigued by the discovery of the shells of sea creatures at the top of these cliffs, a long way above the waves. Most people in those days, if they thought about such things at all, assumed that this must be something to do with the biblical flood. But even though he was the son of a curate, Hooke had doubts, which developed over the years into ideas that culminated in a series of lectures at Gresham College, published posthumously as *A Discourse on Earthquakes*. Way ahead of his time, as we shall see, Hooke realised that the landscape we see around us today is a result of geological processes operating over immense spans of time, far longer than the then popular biblical timescale of Bishop James Ussher. Much later, he recalled how as a child he had observed a cliff made of layers of material, one of which, far above he sea, was a band of sand 'filled with a great variety of Shells, such as Oysters, Limpits, and several sorts of Periwinkles.'*

These activities took place against the background of the Civil War (actually a series of wars), which lasted from 1642 to 1651. Although the Isle of Wight was staunchly Royalist, its geographical isolation just off the south coast of England, and a judicious surrender to Parliament at the beginning of the conflict, spared it from the turmoil suffered by much of the country, but it was a natural place for Charles I to set up a Royalist base when he escaped from Parliamentary captivity in November 1647 (it is widely thought that he was allowed to escape by the Parliamentarians, at a loss to know what to do with him, in the hope that he would flee to permanent exile in France). This adventure came to nothing, but must have made an impression on Robert, who remained a Royalist throughout his life.

All the model-making and wandering abroad in the countryside came to an end, however, in October 1648, when Hooke's father

---

* Quoted by Lisa Jardine.

died. Robert was just thirteen. John had been ill for some time, and knowing that his time was short had made careful provisions for the family. He left the boy as his share 'forty pounds of lawful English money, the great and best-joined chest, and all my books'; there was an additional legacy of £10, which had been held by John in trust for Robert, from the will of Robert's maternal grandmother. The total sum of £50 sounds modest today, and some accounts describe the boy as an impoverished orphan. But in terms of spending power, it was equivalent to about £20,000 today, certainly enough to give him a start in life, even if he would soon have to find a way to earn a living. It may be significant that Robert's inheritance was entirely portable – as Lisa Jardine put it: 'cash, books and a chest to carry them in'. Clearly Robert's future away from the island was already planned. The first step down the road to that future took him as an apprentice to the studio of the portrait painter Peter Lely at Covent Garden in London,* just about at the time the King's adventure on the island came to an end and he was carried off once again, this time permanently, by the forces of Parliament. Charles was beheaded on 30 January 1649.

Hooke was almost certainly introduced to Lely by John Hoskins, who may have been the person who took him from the Isle of Wight to London. It is easy to imagine the likely fate today of a thirteen-year-old boy with £20,000 in his pocket, installed as an apprentice to an artist in Covent Garden, part of the expanding metropolis of London, then home to some four hundred thousand people. But children were expected to grow up more quickly in the seventeenth century, and Hooke, as he soon demonstrated, was no ordinary child, even by the standards of his day.

But Robert did not stay with Lely for long. Almost as soon as he was installed in Lely's studio, Robert had second thoughts. According to John Aubrey, who later became a close friend of Hooke, he decided that Lely had nothing to teach him: he 'quickly perceived what was to be donne, so, thought he, why cannot I

---

* Lely later painted the famous portrait of Oliver Cromwell with 'warts and all'.

doe this by my selfe and keep my hundred pounds?'* According to Waller, Hooke was put off a career as an artist by the smell of the painting materials, which brought on a recurrence of the headaches that had plagued his childhood. Both accounts may, of course, contain part of the truth. And in the light of what happened next, there may be a third thread to the story.

After a brief time with Lely, Hooke enrolled at the prestigious Westminster School, where the headmaster, Richard Busby, held on to his post in spite of his Royalist sympathies and the proximity of Parliament. It is easy to identify the connection that took him there. Cardell Goodman, the rector at Freshwater, had been a pupil at the school, and was a witness to and executor of the will of John Hooke. Our own speculation is that Robert was supposed to be going to Westminster School all along, with his money and chest full of books, but was briefly tempted by the thought of becoming an artist. It is fortunate for the development of science in Britain that he quickly came to his senses and followed what was probably his father's plan.

Busby was an enlightened headmaster (in some ways; he was also a strict disciplinarian) who charged pupils according to their intellectual ability as well as their ability to pay. Some paid as much as £30 a year, which would soon have eaten up Robert's inheritance. But some paid nothing at all, and were lodged in Busby's house. There is no record of what, if anything, Robert paid for his education, but he was one of Busby's special cases, bright but relatively poor boys who did not necessarily follow the regular curriculum (which still concentrated on the Classics, Greek and Latin literature) but had freedom to develop other skills that might be useful in later life. The 'regular' pupils, sons of gentlemen all, and including John Locke, Christopher Wren (three years Hooke's senior, who became his close friend in Oxford) and John Dryden, had no need to get their hands dirty in this way. But it suited Hooke perfectly.

Although he was not often seen at lessons (at least, according

---

* Actually £50, which gives you some idea of Aubrey's reliability.

to Aubrey), during his time at Westminster Hooke mastered Latin and could converse in the language, and studied Greek and Hebrew, like the classical scholars. He also, though, learned to play the organ, a skill that would soon come in handy, and mastered the mathematical works of Euclid. According to Waller:

> he fell seriously upon the study of the Mathematicks, the Dr. [Busby] encouraging him therein. and allowing him particular time for that purpose. In this he took the most regular Method, and first made himself Master of Euclid's Elements, and thence proceeded orderly from that sure Basis to the other parts of the Mathematicks, and thereafter to the application thereof to Mechanicks, his first and last Mistress.

Instead of his lessons, he could be found in one of the workshops associated with the school, where he spent the long hours bent over a lathe that he thought produced his stoop. It seems more likely, however, that he suffered from a condition known as Scheuermann's kyphosis, a curvature of the spine that develops in adolescence and may have a genetic basis but has been linked to poor diet when young.

Hooke's interest in 'Mechanicks' while at Westminster led him, among other things, to devise 'thirty severall wayes of Flying', he later told Aubrey. John Wilkins, the Warden of Wadham College in Oxford, was another person interested in mechanical devices, and had written a book about them, published in 1648, with the splendid title *Mathematicall Magick, or the wonders that can be performed by mechanical geometry*. The book dealt with the use of levers, pulleys and other mechanical aids for practical uses, then went on to more speculative discussion of mechanical automata, including flying machines (ten years earlier, Wilkins had speculated in print about the possibility of flying to the Moon). It seems that Hooke's interest in mechanical devices, and in particular flying machines, was reported to Wilkins by Busby, helping to smooth Hooke's path when in due course he too moved on from Westminster to Christ Church. Indeed, Wilkins gave a copy of his book to the boy while he was still at Westminster and Hooke still had the book at the time of his death. When he made the move to Oxford, he left behind someone who had become a firm friend, not just his schoolmaster. Busby and Hooke remained friends for the rest of Busby's life (he died in 1695), and Hooke was the architect for a church and vicarage built for Busby at Willen, in Buckinghamshire, in the 1680s. When Busby was Archdeacon of Westminster, Hooke carried out several commissions at the Abbey, including repaving the choir, where the black and white marble flooring he had installed can still be seen. But an architectural career lay far in the future when Hooke went up to Oxford in 1653, at the age of eighteen.

The path from Westminster to Christ Church was a well-trodden one. Each year, four Westminster students were awarded scholarships to the college; but Hooke was not one of the four selected in 1653. Instead, he was awarded a choral scholarship, thanks to his musical ability. This seems to have been literally money for nothing, because during the Parliamentary Interregnum such frivolities as church music were banned. In addition, we are told that Hooke acted as a servitor (or 'subsizar') to a 'Mr Goodman'. The position of servitor, acting as a servant to a more wealthy student, was a way for less well off but academically gifted students to make their way at Oxford or Cambridge in those days. The duties might be very light or more onerous, depending on who was being 'served'. But there is no record of a student called Goodman in Christ Church at the time Hooke was up in Oxford. The logical conclusion is that he was being supported by Cardell Goodman, himself a former Westminster scholar and Christ Church graduate, perhaps with the notional title of servitor for administrative reasons. Although Goodman died in 1653, he could well have left money for the purpose. If so, once again it was money for nothing, and a clear indication of the high academic reputation Hooke had already achieved at the age of eighteen.

Hooke's time as a student in Oxford was distinctly out of the usual path of other students. Although he went up to Christ Church in 1653, he did not matriculate (in effect, register to study for a degree) until 1658, and he never took the BA examination, although he was awarded an MA in any case in 1663, after he had left Oxford (this is not, as we shall see, totally unlike what later happened to Edmond Halley). Instead of following a conventional course of study, alongside what (if anything) he was studying in college he worked as an assistant to two of the pioneering scientists of the time, first Thomas Willis and then Robert Boyle.* The connection with Willis, and through him a group of scientists, had begun by 1655, when Hooke was twenty.

At the time, a new way of investigating the world was being pioneered, and its key feature was experiment. Since the time of the Ancient Greeks, philosophers had developed their ideas by logic and reason, without actually getting their hands dirty by carrying out experiments. This led to the wide dissemination of such ideas as the notion that a heavy object falls more quickly than a light object, even though a simple experiment was sufficient to prove the idea wrong. By the early seventeenth century, individual scientists were applying the experimental method – Galileo most famously, who, although he never did drop objects from the tower in Pisa, did do experiments rolling balls down inclined slopes to see what really happened to them. In England, William Gilbert, a physician at the court of Queen Elizabeth, carried out many experiments with magnets and made huge advances in understanding the nature of magnetism, but equally significantly he explained the importance of the scientific method of testing ideas by experiment. Indeed, his writing directly influenced Galileo, who read Gilbert's book *De Magnete*. Another pioneer of the experimental method was William Harvey, who discovered the circulation of the blood. Harvey had an Oxford connection – as one of the King's physicians he had been residing in the city with Charles when the King made it his capital during the Civil War. Ironically, though, the person who had the most direct influence on the new experimenters was not an experimenter himself. Francis Bacon, one of the key politicians of the Elizabethan age, published his ideas about the experimental method

---

* The term 'scientist' only became common much later; they regarded themselves as 'natural philosophers'. But we use the modern term since they were indeed what we now call scientists.

of scientific research in 1620, under the title *Novum Organum*. In essence, Bacon's argument was that progress should be made by collecting facts, forming hypotheses based on study of these facts, then (crucially) using these hypotheses to make predictions that could be tested by carrying out experiments. As long as the experiments agreed with the predictions, the hypothesis being tested could be elevated to the status of a theory, but any theory could potentially be brought crashing down by a single experiment that gave results that did not match its predictions. In due course, the founders of the Royal Society, led by the same John Wilkins we have already met, would explicitly found their institution on their interpretation of Baconian philosophy. And some of those founders were experimenting in Oxford in the 1650s.

The first stirrings of the scientific debates that led to the founding of the Royal Society took place in London, in the mid-1640s, where a group of men, including Wilkins, used to meet to discuss 'experimental philosophy'. This was at the height of the political and religious turmoil of the Civil Wars, and these gentlemen consciously made a decision not to discuss those contentious topics, but to stick with what we now call science – which must have been something of a relief to them from the uncertainties of everyday life. But at that time the group was essentially a talking shop, not a centre for experiments. After the success of the Parliamentary forces, the former Royalist stronghold of Oxford was reorganised, with many people regarded as King's men ejected from their posts and being replaced. This took several of the London group to Oxford, where Wilkins became Warden of Wadham College in 1648, and a member of the triumvirate overseeing the University on behalf of Oliver Cromwell in 1652. In 1656, Wilkins married Cromwell's widowed sister, Robina, who was a couple of decades older than him, cementing his position in the establishment. By then, Cromwell was the Lord Protector, and gave Wilkins a special dispensation to marry even though his post as Warden officially required him to remain celibate. This seems not to have been pure self-interest on Wilkins' part, because John Evelyn, who knew Wilkins well, tells us that he was:

A most obliging person, [who] had married the Protector's sister, to preserve the Universities from the ignorant Sacrilegious Commander and soldiers, who would fain have been demolishing all bothe [Oxford and Cambridge] and persons that pretended to learning.

By the time Hooke came to Oxford, the group of experimental philosophers was already holding regular meetings (sometimes referred to as a 'philosophical club') at Wilkins' rooms in Wadham. It was at this time that they began to put the 'experiment' into experimental philosophy. Hooke was, of course, already known to Wilkins through Busby, and Thomas Willis was another member of the 'club', which was some thirty strong. Willis was a physician and chemist who was particularly interested in the workings of the brain. He was also a member of another 'club' – the Westminster/ Christ Church old-boy network (indeed, he had been a contemporary of Busby at Christ Church as an undergraduate). So it is no surprise that Hooke became an assistant to Willis, living in his house, Beam Hall, opposite Merton College Chapel, and preparing the medicines for Willis' patients, as well as helping out with chemical experiments. It was from Willis, too, that Hooke learned dissection.

If this meant that Hooke was neglecting his formal studies, it certainly did him no harm. And he was certainly more interested in learning, by whatever means, than many of the 'young gentlemen' who regarded their time at university as something of a holiday. Even at the height of the Puritan regime, with daily prayers at 5 a.m. and 5 p.m., services on Thursdays and Sundays and other devotions, it was necessary for the authorities to instruct the Dean of Christ Church to 'take special care to reform all scandalous fashions of long and powdered hair, and habits contrary to the status of the University and that decency and modesty which is necessary for young students', followed by a demand 'to punish the abuse of swearing'. In 1653, the year Hooke went up, another edict took steps 'for the repressing the immoderate expenses of youth in the College, that no gentleman commoner shall battel in the buttery above 5 shillings weekly'. Not that these financial restrictions would have meant much to the impoverished Hooke.

There was, however, one new temptation that Hooke fell for, and consumed eagerly throughout his life. The first record of coffee being brewed in England comes from the diary of John Evelyn, who wrote on 10 May 1637 'There came in my time to the College [Balliol] one Nathaniel, out of Greece . . . He was the first I ever saw drink coffee.' Nathaniel was later sent down (expelled), we don't know why, but went on to become Bishop of Smyrna, so whatever misdemeanour it was didn't harm his career. Perhaps partly thanks to his example, in 1651 the first coffee house in England was opened on the site of what is now The Grand Café, on the High Street. Its proprietor was a man called Jacob, from the Lebanon; the first coffee shop in London was opened the following year, by Pasqua Rosee, from Turkey, in St Michael's Alley, off Cornhill. By the end of the 1650s, there were more than eighty coffee houses in the City of London. Apart from Hooke's personal addiction to coffee (which may help to explain both the long hours he worked and some of his later ailments), this was an important event for science, as well as society at large, because coffee houses became the preferred meeting places of natural philosophers such as Hooke, Halley and their friend Christopher Wren. A coffee house even comes into the story of the discovery of the inverse square law of gravity.

Although 'only' Willis' assistant, Hooke attended meetings of the philosophical club, and absorbed knowledge from its other members, notably Seth Ward, the Savilian Professor of Astronomy; at Ward's request, Hooke devised a mechanism to improve the regularity of a pendulum clock for astronomical timekeeping, and this led to a lifelong interest in clocks and the problem of finding longitude at sea. It was also here that he met Wren, and during his time in Oxford he continued his interest in flying. But the single most important thing that happened to Hooke in Oxford was that Wilkins introduced him to Robert Boyle, with the recommendation, which Boyle accepted, that Hooke should become Boyle's assistant.

Boyle had reached Oxford by a circuitous route. Many accounts simply describe him as a rich aristocrat who had the time and money to indulge his interest in experimental philosophy. But things were never that simple in the England (and, especially in Boyle's case, Ireland) of the middle decades of the seventeenth century. Boyle's father, Richard Boyle, was indeed the Earl of Cork and filthy rich, but he was not the latest member of a long aristocratic line. Richard Boyle was what might now be called an entrepreneur, in the pejorative sense of the term. Born in England in 1566, into a respectable but unremarkable family, he became a penniless orphan before he was twenty and went to Ireland (then an English colony) to make his fortune. With the aid of marriages to a wealthy widow, and after she died to the daughter of the Secretary of State for Ireland, and financial dealings that were often on the shady side of legality, he succeeded in his aim so well that he became possibly the richest man in either Ireland or England, able to buy his title and the respectability that went with it.

Wheeling and dealing didn't take up all of Richard Boyle's time. Along the way he fathered seven daughters and six sons, before the late addition of Robert, on 25 January 1627, when Richard Boyle was sixty and his wife Margaret forty years old. Yet another daughter was born three years later, but complications associated with the birth killed Margaret. The last girl was named after her.

With no mother from the age of three, and far down the pecking order for any inheritance of either titles or money, the Honourable Robert Boyle (to give him the only title due to him) was initially brought up and educated at home, in the care of family retainers, but later went to Eton. There he was recognised as an outstanding scholar at a very young age, and first encountered the books of Nicolaus Copernicus and William Gilbert. But at the age of twelve, in 1639, Robert was plucked out of school and sent with his brother Francis, then fifteen, on the Grand Tour of Europe that was *de rigueur* for the sons of wealthy gentlemen. Their education was not forgotten. They were accompanied by a tutor, and visited many seats of learning – they were in Florence in 1642 when Galileo died. But when they were about to return home, rebellion broke out in Ireland. This was one of the early precursors to the Civil Wars, and although Francis was considered old enough to be summoned home to help suppress the rebellion, Robert was told to keep away until the fighting was over. The rebels, however, were not suppressed without serious consequences for the Boyle family. Two of Robert's brothers (not Francis) were killed, and the grand old Earl of Cork lost most of his money and land. He died in 1643, soon after this phase of fighting finished. So when Robert returned to England in 1644, he had no money and had not been educated for any kind of useful career. Worse, by then the 'proper' Civil War was raging. He was saved by his sister Katherine, thirteen years older than Robert, who had married to become Viscountess Ranelagh, and lived in London but apart from her husband.

At first, Robert lived in Katherine's house. She was a known Parliamentarian sympathiser with many powerful friends in London, which was controlled by Parliament. Robert judiciously never gave any indication of preferring one side or the other in the Civil Wars, probably because he genuinely just wanted to be left alone to get on with his life in peace. Katherine helped him to find a retreat from the turmoil of the times. Their father had left to Robert a small estate in Dorset – not much compared with the large estates in Ireland once intended for the older brothers, but enough for the youngest son, and by chance one of the few possessions the Earl had left at his death. Thanks to Katherine's connections, the estate was not confiscated by Parliament, and Robert was allowed to live there from 1645 onwards, setting up his own laboratory where he carried out chemical experiments. On visits to London, he stayed with Lady Ranelagh, and like-minded experimental philosophers used to gather to meet with him at her house. Boyle referred to this as an 'invisible college'; we don't know who was involved, but there must have been considerable overlap with the group we mentioned earlier, including Wilkins.

Boyle's fortunes improved in the 1650s, after the Civil Wars ended. One of his surviving brothers, now Lord Broghill, was in favour with Parliament for his part in crushing the Irish. This rubbed off on the rest of the Boyle family, and Robert was able to visit Ireland to pick up some of the threads of their former life. Horrified by the terrible conditions of the Irish people, after some soul-searching he got the estates running as a benevolent landlord (by the standards of the time) who used much of the income for charitable ends. This still left him enough to live on comfortably and continue in the role of gentleman scientist back in Dorset. But John Wilkins, who had met Boyle in London and knew his abilities, invited Robert to move to Oxford, where he could not just carry out his own experiments but be in the company of other people with similar interests. After mulling the offer over, Boyle made the move in 1655. He was never formally part of the university (as he put it himself, 'never a Professor of Philosophy, nor a Gown-man'), but he had his own laboratory, and he also (as Wilkins had probably hoped) helped to finance the work of some of his fellow experimental philosophers. Boyle lived and worked in a house known as Deep Hall, on the High (convenient for the coffee shops!), and it was here, in 1656 (the year, incidentally, that Edmond Halley was born), that Robert Hooke came to live and work as Boyle's paid assistant, although possibly they had already met.

Back in September 1653, when Wilkins was already trying to persuade Boyle to move to Oxford, he had sent a letter to him by messenger. It read, in part:

> This bearer is the young man I recommended to you. I am apt to believe, that upon trial you will approve of him. But if it should happen otherwise, it is my desire he be returned, it not being so much to prefer him, as to serve you.*

Lisa Jardine has suggested that the young man in question was Hooke, and that he was sent as part of the attempt to entice Boyle to Oxford, by showing that a skilled assistant would be available there:

> If it be not, Sir, prejudicial to your other affairs, I should exceedingly rejoice in your being stayed in *England* this

---

* *Correspondence of Robert Boyle*, ed. M. Hunter, A. Clericuzo & L. Principe, Pickering & Chatto, London, 2002.

winter, and the advantage of your conversation at *Oxford*, where you will be a means to quicken and direct us in our enquiries . . . shall be most ready to provide the best accommodation for you, that this place will afford.

The immediate plan to persuade Boyle to Oxford that winter was aborted because he had to travel to Ireland to deal with the urgent business concerning the family estates we have already mentioned. The young man, presumed to be Hooke, returned to Oxford. But when Boyle did make the move some two years later, it seems that he was already aware of the abilities of the man who did indeed become his assistant.

The greatest achievement of the Boyle–Hooke collaboration was an improved air pump, which made it possible for them to carry out experiments both at greatly reduced air pressure and at pressures greater than ordinary atmospheric pressure. That simple sentence, though, needs unpacking in order to put Hooke's achievements, in particular, into perspective.

First, although Hooke was a paid assistant to Boyle, this was a genuine collaboration. Hooke was more than a 'mere' technician who did things at Boyle's direction. This was a very unusual – indeed, possibly unique – working relationship for the time, but it is made clear in Boyle's published works, where Hooke is regularly mentioned by name as a co-experimenter. Other assistants are not so acknowledged. Secondly, an air pump might not sound like a dramatic invention today. But in the middle of the seventeenth century it was the highest of high-tech scientific equipment, equivalent, in terms of the insights it gave, to CERN's Large Hadron Collider, or the Hubble Space Telescope, today. It was cutting-edge technology, leading to breakthrough science. And the man who made the air pump, and made it work, was Robert Hooke, still in his early twenties. If there had been Nobel Prizes in the seventeenth century, Hooke would have walked away with one, for this achievement alone.

It all started with an experiment carried out by the Italian Evangelista Torricelli (one of Galileo's pupils) in 1644. This seemed to shed light on a puzzle that had vexed philosophers for centuries: was it possible for a vacuum, nothing at all, to exist? One school of thought held that matter must be continuous; a rival hypothesis described matter in terms of tiny particles (atoms) moving through the void (vacuum, or empty space). Torricelli took a glass tube, closed at one end, and filled it with mercury. He then put a finger over the open end, and submerged that end below the surface of a dish of mercury before taking his finger away and raising the closed end of the tube into the vertical. Instead of all the mercury flowing out of the tube, the level dropped only until there was a column nearly thirty inches high standing above the level of the liquid in the dish, with nothing at all in the space above the column. This seemed to be the definitive proof of the reality of the vacuum, and along the way the height of the mercury in the tube was explained as a result of the pressure of the weight of the air pushing down on the surface of the mercury in the dish. Torricelli had invented the barometer, for measuring atmospheric pressure, and similar instruments were soon tested by being carried up mountains, where the lower air pressure meant that the column of mercury was shorter than at sea level. Which suggested that if the air continued to thin out, then above the atmosphere there must be empty space.

Instead of carrying the equipment up a mountain, Boyle wanted to try it out inside a vessel where air could be pumped out to lower the pressure. If he could make a vacuum inside the vessel, the level of mercury in the column would fall as the air was removed, until it would not be supported in a column at all. But first, he needed a way to make a vacuum in the laboratory. This is where Hooke came in. Otto von Guericke, in Saxony, had already made a reasonably efficient air pump, which he had used to suck air out of two large copper hemispheres that were placed together rim to rim to make a sphere, but with no mechanical fastenings at the join. With air pressure inside the sphere reduced, the pressure of the atmosphere outside squeezed the hemispheres together so tightly that in a famous demonstration made to Emperor Ferdinand III in 1654 thirty horses could not pull them apart.

Von Guericke's pump was large and cumbersome, needing two

men to operate, and, of course, there was no way to see inside his copper sphere. Boyle needed something that could be operated by one man, with a chamber made of glass through which experiments could be observed. He first approached the greatest scientific instrument maker of the time, Ralph Greatorex, in London. But his forte was making precision instruments, and his attempt at the heavier machinery required for the pump was not up to Boyle's needs. So it was Hooke, at the end of the 1650s, who designed and built the breakthrough instrument, using funds supplied by Boyle. He went to London to oversee the manufacture of the heavy components in the workshops there (we don't know if he worked on these himself), then had them taken to Oxford, where he put the pump together and made it work.

The vacuum chamber consisted of a glass sphere fifteen inches in diameter, known as the 'receiver', with a brass lid four inches in diameter, which could be opened to place apparatus inside the sphere. A tapering hole in the base of the sphere stood on top of a tight-fitting brass cylinder, sealed with a leather collar. The brass lid had a small tight-fitting stopper, sealed with oil (referred to by Hooke as 'sallad oil'), that could be turned to tug a string attached to the stopper in order to set off an experiment inside the globe. The cylinder below the globe connected to a brass pump fitted with an ingenious rack-and-pinion system, which allowed air easily to be pumped out of or into the globe. Hooke's pump sucked air from the cylinder using a piston that was connected to a rod cut with teeth which engaged with a gear wheel that could be wound with a handle to push the piston up, forcing air out through a one-way valve, then pull the piston down, leaving a vacuum in the tube. The piston could be pumped up and down repeatedly, sucking more and more air out of the glass vessel. This apparatus became known as 'Boyle's air pump', which it was in the sense that he paid for it and owned it (just as Dolly Parton's hair is her own). But as Boyle acknowledged, it was made by Hooke, and Hooke was the experimenter who operated it during the many investigations that followed. In the fragment of autobiography quoted by Waller, Hooke said:

In 1658, or 9, I contriv'd and perfected the Air-pump for
Mr *Boyle*, having first seen a Contrivance for that purpose
made for the same honourable Person by Mr *Gratorix*, which
was too gross to perform any great matter.

Some idea of the significance of the pump is that, even by the
end of the 1660s, there were only half a dozen comparable air
pumps in Europe, and three of them had been made by Hooke.

Boyle and Hooke carried out many experiments with their
pump and vacuum chamber – Boyle later described forty-three of
them in his book *New Experiments Physico-Mechanical Touching
the Spring of the Air*, published in 1660. These included burning
(or attempting to burn) substances such as candles, coal, charcoal
and gunpowder in a vacuum, with results that convinced them
that fire was not one of the 'four elements' (fire, earth, air and
water) as the Ancient Greeks had taught, but involved a chemical
process. Candles, for example, went out when air was removed
from the globe, and burning coals died away, but, crucially,
reignited when air was let back in. One of the other experiments
showed that water boils at a lower temperature when the air
pressure is reduced. But one of their most important discoveries
is hinted at in the title of Boyle's book. Every stroke of the handle
of Hooke's air pump demonstrated the 'spring' of the air, just like
the springiness felt when using a bicycle pump today, and Hooke
set out to measure this springiness – what we now call air pressure.

Around this time, at the end of the 1650s, the Englishman Richard
Towneley was carrying out experiments with a Torricelli barometer
on Pendle Hill, in Lancashire. He was following the example of
continental experimenters, notably Florin Périer. Like them, he
found that the pressure of the air measured by the barometer is
lower at higher altitude, and he surmised (without carrying out
experiments to test the idea) that the pressure is less because the
air is thinner – that is, less dense – at higher altitude. He mentioned
this idea to Boyle, who asked Hooke to devise a way to test it.

Hooke did this in 1660 or 1661, using a long glass tube shaped
like the letter J, with the top open and the short arm of the J at

the bottom sealed. He poured a little mercury into the top of the tube so that it partly filled the U-bend at the bottom but left some air trapped in the closed end. With the level of mercury the same on both sides of the U-bend, the trapped air was at atmospheric pressure. But Hooke could increase the pressure on the trapped air by pouring more mercury in, forcing some of it round the bend and squeezing the trapped air into a smaller volume. Boyle was short-sighted and bad at arithmetic, so we know for sure that it was Hooke who not only designed the experiment but also made the careful observations and records that showed that the volume of the trapped air was inversely proportional to the pressure applied. Double the pressure, and the volume halves; triple the pressure and the volume is reduced to one-third, and so on. These results were published in the second edition of Boyle's book, in 1662, and became known as 'Boyle's Law', although he did not use that name himself. Hooke's own account appeared in his book *Micrographia*, published in 1665:

> Having lately heard of Mr. *Townly's Hypothesis*, I shaped my course in such sort, as would be most convenient for the examination of that *Hypothesis*.

After describing the experiment (Hooke tells us that the long arm of the J-tube was about fifty inches long), he concludes:

> and by making several other tryals, in several other degrees of condensation [compression] of the Air, I found them exactly answer the former *Hypothesis*.

The discovery itself was significant. The measurements of the springiness of the air fed into the development of theoretical ideas about the nature of matter, leading up to the idea of atoms and molecules flying about in the vacuum and colliding with one another. It also had practical implications, because the idea of making vacuums using pistons, and using the weight of air (atmospheric pressure) to compress pistons, found applications in steam engines. But from our point of view the most important thing about these experiments is the way they were carried out and reported. For the first time, experimental philosophers described their experiments in great detail, along with the way they overcame difficulties and how they interpreted their results. They not only gave a table showing the actual measurements of pressure made in the course of the investigation, but also included alongside these the numbers corresponding to 'What the pressure should be according to the *Hypothesis*'. The match was not perfect; of course there were experimental errors. But they (or rather Hooke) had found that the accuracy of the hypothesis was confirmed within the limits of experimental error. And everything was laid out carefully so that other experimenters could repeat the whole process and see if their results agreed. It was only later, when many other experiments had indeed confirmed this, that the hypothesis was elevated to the status of a law, albeit with the wrong name attached to it.

While in Oxford, Hooke also developed his interests in astronomy and timekeeping, which we have already mentioned. Some of his other activities can wait until we discuss the contents of *Micrographia*. But there was one interest in particular that Hooke at first eagerly investigated in Oxford but then (for sound scientific reasons) abandoned – flying. This change of heart is described in the autobiography:

I contriv'd and made many trials about the Art of flying in the Air, and moving very swift on the Land and Water, of which I shew'd several Designs to Dr. *Wilkins* then *Warden of Wadham College*, and at the same time made a Module [model], which, by the help of Springs and Wings, rais'd and sustain'd itself in the Air; but finding by my own trials, and afterwards by Calculation, that the Muscles of a Mans Body were not sufficient to do anything considerable of that kind, I apply'd my Mind to contrive a way to make artificial Muscles; divers designs wherefore I shew'd also at the same time to Dr. *Wilkins*, but was in many of my Trials frustrated of my expectations.

The details surrounding one other project which Hooke worked on in the late 1650s and early 1660s are less clear, because for commercial reasons (in the hope, never realised, of making a fortune from his invention) for a long time Hooke kept details of his work on clocks and watches secret, and when he did report them he was inclined to exaggerate his achievement to strengthen his case. Nevertheless, it is quite clear that by about 1658 he was deeply interested in the possibility of designing an accurate time-piece – a chronometer – that would solve the problem of finding longitude at sea. This was of vital importance to an emerging maritime power such as England or their Dutch rivals.

Finding the latitude of a ship at sea was a relatively simple matter of measuring the height of the Sun above the horizon at local noon. But determining longitude was a much more difficult problem. It was clear that the person who solved that problem would certainly become rich as a result – even before the estab-lishment of the famous prize of £20,000 offered for the solution by the British government in 1714. Hooke was always concerned about his financial security, and looked into two ways to tackle the problem. The first was based on the idea of astronomical observations, in particular observations of the moons of Jupiter. The four largest moons (discovered by Galileo in 1610) follow regular, predictable orbits around the giant planet, changing their positions relative to one another like the hands of a heavenly clock. These orbits could be predicted from past observations, even before the discovery of the inverse square law of gravity, so by studying tables of predicted patterns (in particular, eclipses of the moons by Jupiter) and comparing them with observations, a mariner could determine the time at the place where the tables were drawn up (such as the home port, or London) and compare that with the local time. Because of the rotation of the Earth, which takes twenty-four hours to complete a 360-degree rotation, local noon is one hour later for each fifteen degrees west of the home base, and one hour earlier for each fifteen degrees east (360/24 = 15); even Oxford time, by the Sun, is five minutes behind the time at Greenwich, in London. So the difference would tell them how far east or west of home the ship was. This was one of the reasons, in addition to his interest in astronomy and the nature of the Universe (Hooke was interested in everything about how the world worked!), that Hooke devoted a great deal of time to developing improved astronomical observing instruments. But making the required accurate observations from the surface of a ship at sea, pitching and rolling in the waves, was totally impractical.

The other way of working out how far east or west of, say, London you were would be to carry 'London time' around with you, in the form of a clock or watch set before starting out on the voyage. But that would require a chronometer that could keep time to an accuracy of a few seconds over an interval of weeks or months. And, again, it had to do so on a ship being tossed about on the waves.

In the middle of the seventeenth century, revolutionary developments in timekeeping devices were taking place. Earlier clocks, going back to the fourteenth century, were powered by slowly falling weights, connected to the gears and wheels of the mechanism by cords wrapped around a bobbin-like drum. The drum rotated as the weight fell, and the rate at which the weights fell was controlled by a so-called verge escapement, involving a toothed 'crown wheel' which was tugged one step (one tooth) at a time by the pull of the falling weight. When the weight reached its lowest point, it was simply lifted (or wound) back up to keep the clock ticking. These clocks were good for measuring the passage of the hours, provided they were re-set at noon, but did not even measure minutes accurately, let alone the seconds. It was Galileo who realised that the time it takes for a pendulum to complete one swing of its arc depends only on the length of the pendulum, and the Dutch scientist Christiaan Huygens who, in 1656, used this, in conjunction with a traditional verge escapement, to produce the first reasonably accurate pendulum clock. A pendulum 39.1 inches (0.994 m) long takes one second to swing one way, and one second to swing back, at 45 degrees latitude on the surface of the Earth; at one time it was proposed that this length should be used to define the metre (making a metre 39.1 inches), but this was not followed up.* Both Huygens and Hooke set out to improve on these devices, being well aware that no matter how accurate it might be on land, a pendulum clock was hardly the most practical timepiece to have on the heaving decks of a ship.

Hooke's key idea was to replace the regular swing of a pendulum with the regular pulse of a coiled spring, vibrating in and out. He also devised an improved escapement. The spring-driven mechanism would work in a clock, but, equally importantly, could be made small enough to be incorporated in a watch compact enough to be carried in your pocket.

This is where the historical chronology becomes murky. Hooke certainly had the idea for such a watch by 1660 (the year of the Restoration, when Charles II came to the throne). But had he made a watch to this design by then? Hooke, via Waller, tells us that he had:

> Immediately after his *Majesty's* Restoration, Mr. *Boyle* was pleased to acquaint the Lord *Boucher* and Sir *Robert Moray* with it, who advis'd me to get a Patent for the Invention, and propounded very probable ways of making considerable advantage by it. To induce them to a belief of my performance, I shew'd a Pocket-watch, accommodated with a Spring, apply'd to the Arbor of the Balance to regulate the motion thereof . . . this was so well approved of, that Sir *Robert Moray* drew me up the form of a Patent . . . [but] the discouragement I met with in the management of this Affair, made me desist for that time.

The discouragement to which Hooke refers is a proposed clause in the patent that says that if anyone else improved upon the design 'he or they should have the benefit thereof during the term

---

* The metre was defined in 1793 as one ten-millionth of the distance from the equator to the North Pole, but is now defined as the distance travelled by light in a specified fraction – about one three-hundred millionth – of a second. The second is defined in terms of a certain number of periods – about 9 billion – of a particular frequency of radiation from the caesium atom. Hooke would surely have been impressed.

of the Patent, and not I'. It is hardly surprising that Hooke refused to sign away his rights in this way (as he put it, it is easy to add to an existing invention), and there the matter rested until a later dispute, as we shall see, blew up with Huygens.

We know that these events took place – a draught copy of the patent survives. But did they happen in 1660, or a little later? The surviving papers are undated, which doesn't help. Some historians suggest that it was actually in 1663 or 1664, and that Hooke later fudged the dates in order to strengthen his case against Huygens. The most careful analysis of the papers has been carried out by Michael Wright of the Science Museum in London.* He concludes that Hooke probably mentioned the invention to Moray in 1662, and revealed the details a year or two later, with the invention then being developed further in 1664, with a timekeeper completed in the summer of 1666. We shall never know for sure, and at this distance in time the priority doesn't matter. What matters is that Hooke certainly did invent a spring-driven pocket watch, unaided, by the early 1660s, while also working as Boyle's assistant (including the discovery of 'Boyle's Law') and carrying out his own investigations of, among other things, flying, astronomy, and the microscopy that features in the next chapter. Apart from the significance of the watch itself, which was indeed a major development, two points are noteworthy about this story. The first is the way Hooke worked on many projects at once; the second is the connection with the dramatic event of the Restoration. Both would be significant in the next phase of Hooke's career.

---

* See Michael Hunter & Simon Schaffer, *Robert Hooke: New Studies.*

CHAPTER TWO

THE MOST
INGENIOUS BOOK
THAT EVER I READ
IN MY LIFE

At the end of the 1650s, England was once again plunged into political turmoil. Oliver Cromwell died on 3 September 1658, and was succeeded by his son Richard, a less competent administrator unable to cope with a Commonwealth that was already in difficulties, with mounting debts and rival factions. In April 1659 Richard was pushed aside and the army took over, raising the prospect of another civil war. Many people who were in a position to do so, Hooke among them, started to make contingency plans. Hooke's youthful imagination had been caught by the sight of the ships entering and leaving Yarmouth, and he now began to consider life as an adventurer and explorer travelling to the Far East. In May 1659, still not yet twenty-four years old, he read a book, *Itenerario*, written by a Dutch traveller, Jan van Linschoten, and made notes, which survive, about the kind of life he could expect if he followed in van Linschoten's wake. He took particular note of the attractions of China, where 'Schollars are highly esteemed'. But before Hooke's plans could come to fruition – if they were ever more than a pipe dream – in the spring of 1660 Charles II was welcomed back to England, and the monarchy was restored. A wave of optimism swept the country, and Hooke, from the staunchly Royalist Isle of Wight, abandoned his plans to travel and looked forward to a future in England, where he was securely established with Boyle and had a growing reputation among the wider circle of experimental philosophers. He published his first scientific paper (as we would now call it), on capillary action, in 1661. But by then, the centre of experimental philosophy was shifting from Oxford to London.

More precisely, the scientific activity was centred around an institution known as Gresham College, in the City of London (the edifice known as Tower 42 now stands on the site, between Broad Street and Bishopsgate). In Hooke's day, the building on that site was a large Elizabethan mansion, once owned by a wealthy merchant, Thomas Gresham. A range of buildings surrounded a square courtyard roughly a hundred yards across. Gresham had died in 1579, and left the income from his investments to have the house converted into a college and to pay for the appointment of seven 'professors' in perpetuity. The professors would be provided with an income of £50 a year for life, and rooms in the college, in return for giving lectures in their specialist subjects once a week in term time. The specialist subjects chosen by Gresham were law, physic (medicine), divinity, rhetoric, music, chemistry and astronomy. The professors were also required to be celibate, although as we shall see the interpretation of this term was rather loose. The status of these posts has waxed and waned over the years, but there are still Gresham Professors giving lectures, even though they no longer have a college to live in.

Hooke's Oxford friend, Christopher Wren, had become the Gresham Professor of Astronomy in 1657, a post he held until 1661, when he returned to Oxford as Savilian Professor of Astronomy. Other experimental philosophers based in, or visiting, London (and crucially including Wilkins, who had become the Master of Trinity College in Cambridge in 1659, but was ejected when the Royalists returned to power, and was now lodging with a friend in Gray's Inn) used to attend Wren's lectures, and got into the habit of meeting up afterwards to discuss the topics raised and other scientific matters. On 28 November 1660, after one of Wren's lectures, the group decided (clearly by prior arrangement) to formalise these gatherings. A record in the Royal Society archive reads:

> Memorandum November 28, 1660. These persons following according to the custom of most of them, met together at Gresham College to hear Mr Wren's lecture, viz. the Lord Brouncker, Mr Boyle, Mr Bruce, Sir Richard Moray, Sir Paule Neile, Dr Wilkins, Dr Goddard, Dr Petty, Mr Ball, Mr Rooke, Mr Wren. And after the lecture was ended they did according to the usual manner, withdraw for mutual converse.

That 'mutual converse' led to the resolution that they would form an association 'for the promoting of Experimentall Philosophy' and:

> That this company would continue their weekly meetings on Wednesday, at 3 of the clock in the term time, at Mr Rooke's chamber at Gresham College; in the vacation at Mr Ball's chamber in the Temple, and towards the defraying of expenses, every one should, at his first admission, pay downe ten shillings and besides engage to pay one shilling weekly . . . Dr Wilkins was appointed to the Chair, Mr Ball to be Treasurer, and Mr Croone, though absent, was named the Registrar.

This was the beginning of the Royal Society, whose members became known as 'Fellows'. Because of Wilkins' reputation as a Parliamentarian, it became politic for him to take a back seat (at least formally), and Sir Robert Moray was installed as President of the fledgling association on 6 March 1661. In no small measure thanks to his skill at political wheeling and dealing, the Society gained its first Royal Charter in 1662, with Brouncker now named as President, but this Charter proved unsatisfactory (for obscure reasons), and was replaced by a second Charter in 1663, formalising the name as 'the Royal Society of London for Promoting Natural Knowledge'.* The Society had a coat of arms, and a motto, *Nullius in Verba*, which can be translated as 'take nobody's word for it'. In other words, carry out experiments and test hypotheses for yourself, do not rely on hearsay. It would be Hooke who soon put that fine sentiment into practice. We shall always refer to the institution as the Royal Society (even for the period before the award of the first charter), the Royal, or the Society; one of the aims of seeking royal status was to get financial support from the King, which was never forthcoming, but the status did encourage rich dilettantes to offer their support, if only by becoming Fellows and (sometimes) paying their subscriptions.

As early as December 1660, the Society laid out the ground rules for doing experiments, and recognised the need for 'curators' who would carry out the experiments. At first, this role was carried out by the most expert Fellows (known as virtuosi), but this was not a success, and it became clear that they needed somebody who could do the job full time. In the early 1660s, Boyle was spending some of his time in Oxford and part at his sister's house in London, where he had a laboratory. Hooke accompanied him and was well known to the Fellows (his little paper on capillary action is mentioned in their records). By 1661, Boyle and Hooke were developing an improved air pump, and Boyle gave their original pump to the Royal, where it languished with nobody able to operate it satisfactorily. This was another indication of the need for a skilled curator who could make things work. And who better than the man who had designed and built that pump?

So on 12 November 1662 Sir Robert Moray proposed, and the Fellows accepted, that Hooke should be appointed Curator of Experiments 'to furnish them every day when they met, with

---

* We have explained the background to the formation of the Royal Society, and the backgrounds of its founders, in our book *The Fellowship* (Allen Lane, 2005).

three or four considerable Experiments', as well as following up topics for investigation suggested by the Fellows. The only snag was, the Royal did not have any funds with which to pay him. The solution was that in effect Boyle 'lent' Hooke to the Royal Society until 1665, when a combination of circumstances (not all of them straightforwardly honest) stabilised the situation.

The Royal had notionally set Hooke's salary as £80 a year, even though they were not paying it. Nor were they able to provide him with accommodation, so he had to make do with temporary lodgings. Partly as compensation, in recognition of his value he was elected as a Fellow of the Royal Society on 5 June 1663, with all the usual fees and subscriptions being waived. The prospect of establishing the relationship on a proper basis came in May 1664, when Isaac Barrow (the successor to the Laurence Rooke in whose rooms the Royal had its early meetings) resigned his post as Gresham Professor of Geometry to become the first Lucasian Professor of Mathematics in Cambridge (where he came across a student called Isaac Newton, who later became the second Lucasian Professor). Before he left for Cambridge, Barrow had been giving some of the astronomy lectures in place of Dr Walter Pope, Wren's successor, who was temporarily away from London. After Barrow left, Hooke took on those temporary duties, and received the appropriate stipend, while Pope was away. Who better to be Barrow's replacement?

There were two candidates for the post: Hooke, who had strong support from the Royal, and a physician, Arthur Dacres. On 20 May 1664, a committee ('The Court') met to decide between them, and duly announced their verdict:

two learned persons viz. Dr Arthur Dacres and Mr Robert Hooke being suited for the same, their petitiones being Read their ample Certificates considered and the matter debated The Court proceeded to election and made thereof the said Dr Dacres to supply the said place of Geometry Reader in the College.

A few days later, perhaps while drowning his sorrows, Hooke bumped into a wealthy merchant, Sir John Cutler, in a public house. He knew Cutler through a mutual friend, and gloomily recounted the tale. Cutler's response was to tell Hooke to cheer up, because he, Cutler, would provide the financial support Hooke needed by creating a post for him to lecture on the History of Trades, at the same remuneration as a Gresham Professor – £50 a year. Before the arrangement could be formalised, however, the Royal Society got wind of some irregularities surrounding the appointment of Dacres. It turned out that the actual committee had voted for Hooke by five to four, but that the Lord Mayor of London, Sir Anthony Bateman, who was present as an observer but not a member of the committee, then voted for Dacres, making a tie, and followed this up by claiming the right to a casting vote in favour of his man. Bateman's term as Lord Mayor came to an end shortly after this fiasco, and he was succeeded by Sir John Lawrence, a more straightforwardly honest man who knew Hooke's abilities. Following formal representations by the Royal, a committee of investigation chaired by Sir John met on 20 March 1665 and concluded:

> that Robert Hooke was the person legally elected and accordingly ought to enjoy the same with the Lodgings profits and all accommodation to the place of Geometry Reader appertaining.

In the months before the appeal was heard, the Royal acted with underhand cunning to secure the benefits of Cutler's offer for themselves. On 27 July 1664, the Council of the Royal formally voted to appoint Hooke as Curator of Experiments with a salary of £80 a year, but kept this secret while they negotiated 'on Hooke's behalf' with Cutler. It was agreed that Hooke would give what became known as the Cutlerian Lectures, on practical applications of science 'to the advancement of art and nature' but on specific topics chosen by the Royal. And Cutler's money would be funnelled to Hooke through the Royal. So when Hooke was formally appointed as Curator on 11 January 1665, the Royal only had to add £30 a year for his income to be made up to the promised £80. The situation was compounded when Cutler (possibly piqued by this, or maybe just unreliable) failed to pay his share most of the time, leading to tedious legal hassles only resolved in Hooke's favour after Cutler's death, in 1696 (for the first ten years, the Royal also had trouble finding the money to pay their contribution to his salary). But still, as he did get the Gresham chair Hooke was reasonably comfortable from the time he was installed as Gresham Professor in March 1665 (he had actually been lodging in rooms in the College since the previous September). As well as the income, he had a parlour, library and two smaller rooms in a first-floor apartment, a workshop on the ground floor, cellar rooms providing further space for his experimental work, and a garret for a servant. He was able to keep at least one servant, usually a girl, and usually on more than friendly terms, as we discuss later. He was a gregarious and friendly man (at least until old age and infirmity made him more grumpy), who welcomed visitors to his home, as well as mingling with his friends in the coffee houses. At the age of twenty-nine, he was settled for life, with no need of patronage.

Hooke was a diligent lecturer, unlike many of his fellow Gresham Professors. Some didn't even live at the College, but let out their rooms and enjoyed a quiet life in the country, or even in another country. Hooke's duties (in addition to his work for the Royal, remember!) were to give his lectures on Thursdays in term time,* in Latin between 8 a.m. and 9 a.m. and the same lecture in English between 2 p.m. and 3 p.m. He seems to have always had the lectures prepared and been available to do his duty, but very often, as his diary records, nobody turned up to listen to them. He also gave the Cutlerian Lectures, officially during the vacations but sometimes on other occasions; many of these were collected and published in 1679. These wandered far

---

* These were the terms adhered to by the legal profession, which evolved into the academic terms.

from the original brief, which makes them much more interesting to us even if it helps to explain Cutler's reluctance to pay Hooke.

But that is getting ahead of our story.

The year 1665 was a turning point for Hooke in other ways, but before we discuss the changes in his life that took place in the second half of the 1660s, we should go back to look at his scientific achievements in the first half of that decade.

Some idea of the breadth of Hooke's activities can be gleaned from a 'wish list' he wrote at the beginning of the 1660s of the projects he had in mind:

Theory of Motion:
    of Light
    of Gravity
    of Magneticks
    of Gunpowder
    of the Heavens

Improving shipping
    – watches
    – Opticks
    – Engines for trade
    – Engines for carriage

Inquiry into the figures of Bodys
    – qualitys of Bodys

Hooke worked on many of these projects (and others) in parallel.

We can only pick out the highlights, and describe them consecutively, even when two or more of them overlapped chronologically. The extraordinary fact is, though, that Hooke worked on an array of subjects at the same time, while also giving his lectures and doing more experiments at the behest of the Royal Society. But let's begin with some of his first work for the Royal, using the air pump that Boyle had given to the Society, and which only Hooke could operate effectively. With that tool, he carried out the two duties that were the key to the survival of the Royal Society, a survival that he alone ensured. First, he entertained the Fellows with dramatic demonstrations. The importance of this cannot be over-emphasised. It was this kind of showy demonstration that fascinated the more dilettante Fellows and which brought in a flow of subscriptions to keep the Royal afloat, even if that flow was sometimes only a trickle. Secondly, and much more important to us, he carried out experiments that advanced scientific knowledge profoundly.

A good example of Hooke's skill as a showman, and the way this linked up with his scientific studies, is provided by his work with hollow glass balls. He delighted his audience with demonstrations in which the balls 'exploded' as they cooled down after being blown from molten glass, and the way air rushed into them when they were placed under pressure in the chamber (receiver) of the vacuum apparatus and cracked open. Among other things, though, this set Hooke thinking about the strength of arches and other curved structures, so the experiments fed directly into his later work as an architect.

It also seems that Hooke was not afraid to experiment on himself. In his diary entry for 7 May 1662, John Evelyn (himself a Fellow) describes a meeting of the Royal Society attended by the King's cousin, Prince Rupert:

> I waited on Prince Rupert to our Assembly, where were tried several experiments of Mr. Boyle's Vacuum: a man thrusting in his arme, upon exhaustion of the ayre, had his flesh immediately swelled, so as the bloud was neere breaking the vaines, & insufferable: he drawing it out, we found it all speckled.

There is little doubt that the experimental subject was Hooke himself. Some years later, he built a receiver large enough to sit in, and did so while an assistant pumped the air out. He described how this caused pain in his ears, deafness and giddiness, before he decided enough was enough and the air was let back in. But a discussion of Hooke's most important work with the vacuum pump can wait until we discuss his great book, *Micrographia*.

Although he was not afraid to experiment upon himself, Hooke was far more reluctant than most of his contemporaries to experiment on other animals, at least when it clearly caused them pain. At the beginning of the 1660s, nobody knew exactly what the importance of breathing was in sustaining life. One school of thought held that although the circulation of the blood was clearly important, the role of breathing was simply to act as a pumping mechanism, by which the in and out motion of the thorax stirred up the blood and kept it flowing. The idea that something from the air mixed with blood in the lungs and was essential for life was a minority view. In one indecisive experiment at the beginning of 1663, Hooke placed a live chick and a burning lamp in a sealed chamber to see which one lasted longer. The lamp went out, but the chick survived. This, however, neither proved nor disproved the hypothesis. It was not until November 1664 that Hooke, possibly at Boyle's suggestion, conceived of an experiment on a living dog, which could be dissected 'displaying his whole thorax, too see how long, by blowing air into his lungs, life might be preserved, and whether anything could be discovered concerning the mixture of the air with the blood in the lungs.'

The gruesome experiment was carried out on 7 November. With the dog cut open and all its organs exposed, unable to breathe of its own volition, air was pumped into the lungs of the dog by a pair of bellows through a hollow cane stuck into a hole in the dog's windpipe. The experiment was a success, in that the dog lived during it. As Hooke wrote to Boyle on 10 November 1664:

> at any time, if the bellows were suffered to rest . . the animal would presently begin to die, the lungs falling flaccid, and the convulsive motions immediately seizing the heart and all the other parts of the body; but upon renewing the reciprocal motions of the lungs, the heart would beat again as regularly as before, and the convulsive motions of the limbs would cease.

But in the same letter, Hooke confessed that although the experiment suggested several other lines of investigation:

I shall hardly be induced to make any further trials of this kind, because of the torture of the creature: but certainly the enquiry would be very noble, if we could any way find a way so as to stupefy the creature, as that it might not be sensible [conscious].

Three years later, Hooke was asked to repeat the demonstration, but initially refused. Two doctors, who were less squeamish about such matters, tried to replace him, but made such a mess of the operation that Hooke, by then an employee of the Royal, was ordered to do it and repeated his earlier success.

At the end of 1662 in another series of experiments, he demonstrated how a hollow glass ball that would float on top of cold water gradually sank to the bottom when the water was warmed, or could be made to 'hover' partway up the vessel if the temperature conditions were just right. He correctly suggested that the heat 'loosened' the water (that is, reduced its density), which was another step towards an understanding of matter as made up of atoms and molecules. He also invented (at least in principle; we are not sure if he made it) an efficient water heater in which a heated piece of copper at the bottom of a tub of water would heat the whole vessel as the warm, loosened water rose to the top and was replaced by descending cooler water. He had 'discovered' convection – but he went too far when he speculated that this might make it possible to manufacture a perpetual motion machine in which the water circulated endlessly through a system of pipes without any further heating once it had been started. More practically, he pointed out that because the cold sea at high latitudes could support heavier ships than the 'loosened' water closer to the equator, ships setting out from polar latitudes to the tropics should not be fully laden. Much later, starting in the late nineteenth century, merchant ships were marked with 'Plimsoll lines' showing exactly how far they could be safely loaded, depending on the waters they were visiting.

Hooke's investigations of pressure, density and convection fed directly into another lifelong interest of his: the weather, and the possibility of forecasting the weather. This became a major thread of his work in September 1663, when Wilkins, on behalf of the Royal, asked Hooke to collect daily records of the weather, in the hope that these might reveal patterns that could be used in prediction. Wilkins probably had in mind a simple note of whether it was sunny or cloudy, rainy or dry, and so on. But Hooke never did anything by halves, and he began by setting out a systematic schedule of everything scientific weather observers should take note of (wind speed and direction, temperature, humidity, air pressure, the appearance of the sky, and so on) before he put those principles into practice. He said that the weather observer should also note what illnesses (human and animal) were rife at the time, what diseases and pests were affecting the crops, and many other items. All of this was to be recorded in a standard format, so that the data for each month could be scanned at a glance. Among these details, Hooke was the first person to establish a standard list of terms to describe different kinds of cloud cover.

The project soon developed far beyond the simple record keeping envisaged by Wilkins. You can't keep reliable records unless you have reliable instruments to measure with, and a reliable scale against which to calibrate those measurements. It was Hooke who defined the freezing point of distilled water as the zero of temperature, marked on sealed glass thermometers, an idea enshrined in later temperature scales with the boiling point of water set as the second fixed number, though by then nobody remembered it had been Hooke's idea. He realised that thermometers were affected by the expansion and contraction of the glass as it warmed and cooled, and studied the effect. To measure humidity, he observed the way the ears of the wild oat and wild geranium bent more or less as the humidity changed, and adapted this for use in a hygroscope.* But he made perhaps his most

---

* All of this highlights a key feature of the English scientific revolution. Developments in scientific understanding went hand in hand with developments in scientific technology, as indeed they still do. The key developments in Hooke's day were things like telescopes, microscopes, barometers and thermometers. Hooke had a hand (sometimes both hands) in the development of all these instruments, and many more. For that alone he would have been a key contributor to the revolution.

significant weather discovery in September 1664, just after he first moved into rooms at Gresham College.

This harked back to his work with Boyle on 'Mr. Townly's hypothesis'. It used a portable barometer shaped like a letter J, as in that work, but this time with the long end of the tube closed and the bottom (the short limb of the J) open to the air. Mercury in the U-bend of the J would be pushed down more when the atmospheric pressure was higher, forcing the mercury on the other side further up the long arm of the tube. Similarly, when the pressure fell, the mercury in the long arm fell. By the end of 1663, Hooke had converted this into a 'wheel' barometer, with a pointer that moved around a dial like the face of a clock to show how the pressure was changing. He did this by twisting a thread around the axle of the pointer, with the other end of the thread attached to a weight floating in the mercury in the open end of the tube, and a counterweight on the other side of the axle hanging free in the air. As the mercury moved up and down, the thread tugged the pointer round the dial one way or the other. And if the friction of the axle made it stick, all you had to do was to tap the barometer to get it to unstick and move to the appropriate position.

On 6 October 1664, Hooke wrote to Boyle to tell him of a great discovery he had made using one of these barometers:

I have also, since my settling at *Gresham* college, which has been now full five weeks, constantly observed the baroscopical index . . . and have found it most certainly to predict rainy and cloudy weather, when it falls very low; and dry and clear weather, when it riseth very high, which if it continues to do, as I have hitherto observed it, I hope it will help us one step towards the raising a theoretical pillar, or pyramid, from the top of which, when raised and ascended, we may be able to see the mutations of the weather at some distance before they approach us, and thereby being able to predict, and forewarn, many dangers may be prevented, and the good of mankind very much promoted.

Hooke's vision was not immediately fulfilled: too many other elements, not least rapid communication systems to enable the collation of data from widespread observers, would be required before the vision became reality. It would be two centuries before Admiral Robert FitzRoy 'invented' the weather forecast, but when he did so the kind of links between atmospheric pressure and weather that Hooke had discovered were a key ingredient. And, as FitzRoy's rank highlights, among the 'many dangers' Hooke referred to were the hazards of storms at sea.

Although this particular development was of no immediate benefit to mariners, as we mentioned in connection with Hooke's work on timekeepers, maritime matters were of vital importance to England in the second half of the seventeenth century, and therefore they were of vital importance to the Royal Society as a means of proving its worth to the King. Naval wars with the Dutch involved fleets as far away as America, the Caribbean, West Africa and even the East Indies. It was during a lull in these activities (under the Treaty of Breda, also known as the Peace of 1667) that England formally gained the former Dutch colony of Nieuw Amsterdam, which they had captured in 1664, and promptly renamed it New York. Hooke invented several devices for studying the sea, or working under the waves. One was a depth sounder, which worked by dropping a hollow ball attached to a heavy weight into the sea. When the weight hit bottom, it released the ball, which floated to the surface. By timing how long it took before the ball surfaced, the depth could be calculated. At least, it could in a flat calm with good seeing conditions. In practice, under less than ideal conditions, from the small ships of the seventeenth century the balls could not be spotted as soon as they surfaced (if at all) so the technique was impractical. In the nineteenth century, however, the same idea was dreamed up, independently, by an American, J. M. Brooke, and was used to measure the depth of the sea bed when the first transatlantic telegraph cables were laid in the middle of that century.

Another of Hooke's devices was more immediately successful. This was a bucket on a long line, with hinged lids that allowed it to fill with water at depth, but closed when it was pulled to the surface. This was effective in bringing back samples, which could be studied to measure such things as the saltiness and (with luck) the creatures that lived at depth.

In February 1664 (still before he was being paid by the Royal), Hooke served on a committee that investigated the practical possibilities of diving. He devised a system where a diver working on the bed of a river, or in shallow water at sea, could be supplied with a succession of air-filled lead boxes lowered from the surface, from which he could breathe through a tube. This was reasonably successful during trials in a large tub set up outside the Royal and in the Thames. These and other ideas, including diving goggles, a life jacket, and plans for a submarine, were summed up in an account Hooke published in 1691, but they are only tangentially of interest to our story of Hooke the scientist, as another example of his versatility and capacity for hard work.*

But another aspect of Hooke's maritime work ties in more closely with the main thread of our story. This was his interest in the use of astronomy for navigation, which led him to design and manufacture more accurate instruments for measuring the height of the Sun and stars above the horizon – a key to determining latitude, but also a key to measuring the positions of the stars relative to one another more accurately for other astronomical purposes. This involved better sights (in effect, little telescopes), and instruments calibrated and marked to exquisite precision. One of Hooke's instruments (a quadrant), presented to the Royal in February 1665 (while in the middle of the hassles concerning his appointments as Cutlerian Lecturer and Gresham Professor), was just seventeen inches across, but could measure angular distances as small as ten seconds of arc. Since there are 60 seconds in a minute of arc, 60 minutes in a degree, and 360 degrees in a circle, this means that the instrument could measure precisely angles that are only 1/360th of a degree, or 0.0000077 of a circle.

---

* The interest in undersea exploration would later be shared by Hooke's younger contemporary, Edmond Halley, who was only eight in 1664.

The unprecedented accuracy of Hooke's instruments led to an argument with the much older astronomer Johannes Hevelius of Danzig, who could not believe the superiority of Hooke's designs; the controversy, detailed later, also brought in Edmond Halley, in one of his first missions as a Fellow of the Royal Society.

In much of his astronomical work, especially in the first half of the 1660s, Hooke collaborated with his friend Christopher Wren, who was based in Oxford but still in communication with the Royal. Astronomers of the time were lucky enough to see several comets, and in December 1664 the Royal asked Hooke and Wren to make observations and report on a new comet that had become visible.* Hooke observed from London, Wren from Oxford, and their results plus measurements from other observers were combined and reported by Hooke. Pepys attended a lecture at Gresham College on 1 March 1665 and tells us that on that day (a couple of weeks after he had demonstrated his quadrant), Hooke talked about:

> the late Comett, among other things proving very probably that this is the very same Comett that appeared before in the year 1618, and that in such time probably it will appear again – which is a very new opinion.

New to Pepys, and to Hooke, although we now know that the English clergyman and astronomer Jeremiah Horrocks had speculated along the same lines – that comets follow closed orbits around the Sun – three decades earlier. It happens that Hooke was wrong about this particular comet: it was not the same one that was seen in 1618, and it did not return in 1711. But the improving telescopic technology of the time was starting to show astronomers that comets did not move in straight lines, but followed curved paths through space; this was the beginning of the idea that led Halley, before too long, to make the prediction of the return of the comet that now bears his name. The significance for Hooke's story is that it seems

---

* Hooke and Wren were probably related by marriage. One of Wren's sisters had married a John Hooke, a member of a Hampshire family with relations just over the water on the Isle of Wight. In their correspondence, Wren often addressed Hooke as 'Cousin'.

that, by the mid-1660s at the latest, he was already thinking about the possibility that comets (and therefore the planets) were under the influence of some kind of force, reaching out to them across space from the Sun itself. He realised that comets are part of the Sun's family, not something weird or magical. This was among the insights that led him to carry out several experiments to investigate the nature of gravity, which we describe later. It is worth getting slightly ahead of our story, however, to highlight one of Hooke's most important insights (perhaps *the* most important), which (like so many of his ideas) has been misattributed for hundreds of years.

Going back into the mists of time, it had been assumed by natural philosophers that the 'natural' motion of objects such as planets unaffected by friction or other forces was circular. This had to be so, they reasoned, because circles are perfect, and only perfection could be at work in the heavens. They interpreted the seemingly irregular motion of planets in terms of epicycles, where the planets were constrained to move in small perfect circles around points which themselves moved in perfect circles around the Earth, or the Sun. When, only a short time before Hooke was born, Galileo carried out experiments involving balls rolling down inclined planes, he found that the balls rolled off the end of the ramp horizontally – literally, towards the horizon – and he realised that if there were no friction they would keep rolling for ever. But he knew that the Earth was round, so to him 'horizontal' motion meant always moving towards an always receding horizon, in a circle around the Earth. It was Hooke who realised, partly from his studies of comets, that any object that is not acted upon by an external force will keep moving in a straight line. Does that sound familiar? It should. It is something we all learn in school, where it is called 'Newton's First Law' of motion. But it was Hooke who came up with it, and who (as we shall see) explained it to Newton.

On 21 March 1666, when nobody outside Cambridge and few people inside Cambridge had heard of Isaac Newton, Hooke gave a lecture to the Royal about gravity, where he presented some of these ideas. He described several experiments involving his study of gravity, which he stated was 'one of the most universal active principles in the world' and set out his ambition to determine:

> whether this gravitating or attractive power be inherent in the parts of the earth [and] whether it be magnetical, electrical, or of some other nature distant from either

as well as 'to what distance the gravitating power of the earth acts'.

On 23 May that year he presented his big idea to another meeting of the Royal, and in a paper entitled 'Inflexion of a Direct Motion into a Curve by a Supervening Attractive Principle'. In that lecture (and many times afterward) Hooke used a long pendulum, with the bob moving in a circle, or (crucially, in terms of understanding the motion of the planets) an ellipse, not just to and fro; this demonstrated the nature of orbital motion, which, he pointed out, required a force (in this case, supplied via the string of the pendulum) to keep the bob 'in orbit'. By attaching a secondary, shorter string, with its own bob, partway down the pendulum he could also demonstrate the motion of a 'moon' around a 'planet'. The idea he presented to the Fellows (which really was 'a very new opinion') was that the natural motion of a planet is in a straight line – a tangent to its orbit – and that it is deflected from this tangential path by a force of attraction stemming from the centre of the planetary system – that is, a force emanating from the Sun. As he explained to the Fellows:

> I have often wondered why the planets should move about the Sun according to Copernicus's supposition, being not included inn any solid orbs* . . . nor tied to it, as their centre, by any visible strings.

He stressed that 'all bodies, that have but one single impulse' ought to move in straight lines, and inferred that there must be

---

* A reference to the old idea of 'crystalline spheres' surrounding the Earth and carrying the planets.

another 'impulse' acting on the planets. If that impulse were a force of attraction from the Sun then:

> all the phenomena of the planets seem possible to be explained by the common principle of mechanic motions [and] the phenomena of the comets as well as of the planets may be solved.

These two ideas, 'Newton's' first law and the force of attraction between the Sun and planets (an inward, or centripetal, force), are the keys to the 'Newtonian' revolution in science that took place two decades later. It might have happened sooner, and had a different name, if Hooke's attention had not been diverted by dramatic developments in England in 1665 and 1666. Conveniently for us, however, he had summed up what he described as his 'first endeavours' in a book published just before those changes took place.

Micrographia, Hooke's great book, was written and published on the instructions of the Royal Society as a deliberate attempt to promote the Society and its aims. Hooke has been described as a 'reluctant author',* and almost all of his published work resulted from his contractual obligations, primarily to the Royal Society and to a slightly lesser extent to John Cutler and in connection with his role as a Gresham Professor. But the background to *Micrographia* predates Hooke's appointment as Curator of Experiments.

At the beginning of the 1660s, Christopher Wren was supposed to be preparing a book of microscopical observations for presentation to the King, who had seen some of his drawings of microscopic objects and been impressed by them, but the newly appointed Savilian Professor of Astronomy found that he had too much on his plate, and passed this task on to Hooke, who took over the work in September 1661. The design and manufacture of optical instruments – telescopes and microscopes – was improving dramatically at this time, and although Hooke was involved in

---

* See Michael Hunter's contribution to Bennett et al.

developing some of the ideas that went into these instruments, he relied on expert craftsmen, notably Richard Reeve, for the tools of his trade. As he put it in his book: 'all my ambition is that I may serve to the great Philosophers of this Age, as the makers and grinders of my Glasses did to me'.

By the end of 1662, Hooke was presenting some of his microscopic studies to the Royal. The first of these observations, presented in December that year, dealt with the patterns of ice crystals seen in 'frozen urine, frozen water, and snow'. The Fellows were sufficiently impressed that at the Council meeting of 25 March 1663 Hooke was 'solicited to prosecute his microscopical observations, in order to publish them'. In the months that followed, Hooke made many specific observations at the behest of individual Fellows, as well as following up his own interests. The Council kept a keen eye on the progress of the work, with the book intended to provide an example of the experimental method, which was at the heart of their philosophy, and which they explicitly took from Bacon. In the book, Hooke emphasises the need 'to begin to build anew upon a sure Foundation of Experiments', and explicitly cites the 'Noble and Learned' Bacon as an inspiration. The book was partially intended as propaganda for the Society itself and for the new way of studying the world. It succeeded dramatically on both counts, thanks to Hooke's known genius as a scientist and his perhaps unexpected skill as a writer. But it only got into print after some heart-searching by the Council, which has been detailed by John Harwood.*

Hooke had more or less enough material for his book by March 1664, a year after he had formally been instructed to carry out the work. By then, the Royal had chosen a printer and discussed such details as the official Royal Society imprimatur to go in the front of the book. This emphasised in the clearest way that it was a Royal Society book, stating that:

---

* See his contribution in Hunter & Schaffer.

By the Council of the Royal Society of London for Improving of Natural knowledge.

Ordered, That the Book, written by Robert Hooke, M.A. Fellow of this Society, Entitled, *Micrographia*, or some Physiological Descriptions of Minute Bodies, made by Magnifying Glasses, with Observations and Inquiries thereupon, Be printed by John Martyn and James Allestry, Printers to the said Society

Novem. 23.
1664.   Brouncker. P.R.S

But in the interval from March 1664 to November 1664, the contents of the book had been carefully vetted and discussed by selected Fellows. This caused them some disquiet, because – strictly speaking, exceeding his brief – Hooke did not restrict himself to presenting the observations that he had made with the microscope, but also offered theoretical explanations for why things might be the way they are. He also professes a mechanistic view of Nature, pointing out in the Preface that the reason why we may hope to use mechanical techniques – experimental science – to reveal the workings of the world is that the world operates on the same principles as a machine:

We may perhaps be inabled to discern all the secret workings of Nature, almost in the same manner as we do those that are the productions of Art [artifice], and are manag'd by Wheels, and Engines, and Springs, that were devised by humane Wit.

All of this elevated Hooke's perceived status to that of a natural philosopher, rather than a 'mere' mechanical experimenter. But if his ideas were wrong, the Royal did not want to be seen to endorse them. Ultimately, the Council decided to allow Hooke's speculations to appear in the book, but only if it was made clear that they were his alone, and not the official view of the Society. They ordered:

That the president be desired to sign a licence for the printing
of Mr. HOOKE'S microscopical book: And, That Mr.
HOOKE give notice in the dedication of that work to the
society, that though they have licensed it, yet they own no
theory, nor will be thought to do so: and that the several
hypotheses and theories laid down by him therein, are not
delivered as certainties, but as conjectures; and that he intends
not at all to obtrude or expose them to the world as the
opinion of the society.

Hooke complied, and one result of all this is that we can be sure
the book is all his own work, enhancing his reputation even more.
And he wrote in English, in the first person, making his ideas widely
acceptable. The book was the first scientific best-seller. Samuel Pepys
saw the sheets being prepared when he happened to visit the book-
binders on other business, and promptly ordered a copy of the
book. He received it on 20 January 1665, and the next evening 'sat
up till 2 a-clock in my chamber, reading of Mr. Hooke's Microscopicall
Observations, the most ingenious book that ever I read in my life'.*
A couple of weeks later, Pepys was himself admitted as a Fellow of
the Royal Society, and noted in his diary the luminaries present at
the meeting. 'Above all,' he tells us, 'Mr Boyle today was at the
meeting, and above him Mr Hooke, who is the most, and promises
the least, of any man in the world that I ever saw.' In other words,
in spite of Hooke's unprepossessing appearance, Pepys rated him
above Boyle as a scientist. Clearly, this was at least partly thanks to
the impression made by *Micrographia*.

To us, the speculations that gave the Royal cold feet are more
significant than the illustrations that were the original *raison d'être*
for the book, astonishing though they were at the time, and still
are, considering the difficulties Hooke had to cope with. Remember,
for example, that the only light sources he had were the Sun,
candles and simple oil lamps. In a standard setup, light from an

---

* We have seen it suggested that what impressed Pepys were the illustrations in the book.
But, fine though these are, you don't sit up until 2 a.m. looking at drawings of a louse
or the eye of a fly; what impressed him (and us) were the words.

oil lamp was focused first through a globe containing a transparent solution of brine, and then through a lens on to the specimen he wanted to study. Straining his eyes to concentrate on the image, he then had to draw what he saw with meticulous precision. *Micrographia** contains sixty illustrated 'observations', fifty-seven of them microscopic and three astronomical, made with the aid of a telescope. In a demonstration of his skill as a communicator and his methodical way of working as a scientist, Hooke begins with observation 'of the Point of a sharp small Needle'. 'As in geometry,' he writes, 'the most natural way of beginning is from a Mathematical *point*.' He goes on to describe, with illustrations,† how even the smoothest, sharpest needle looks rough and rounded under the microscope, and he makes a digression to describe the appearance of full stops, both printed and handwritten, which were abundantly 'disfigur'd' even when they appeared perfectly round to the human eye. And he is not averse to a pun, saying after a digression 'But to come again to the point . . . ' The style is easy and accessible even to modern eyes, and the illustrations still stunning. Although in modern times some critics have suggested Hooke could not possibly have seen the detail he claimed, Brian J. Ford, an expert in the history of microscopy, found that by using similar instruments and making careful adjustments of light and focus he could indeed reach the level of detail reported by Hooke. We shall not, however, describe each of the sixty observations in detail. Instead, we shall follow the example of Hooke's biographer Margaret 'Espinasse in picking out four key topics that helped to revolutionise seventeenth-century science.

The first highlight is Hooke's work on light and optics, which is doubly important because it would lead to an intense disagreement with Newton, and one of the most misunderstood comments in the history of science (see Chapter Four). Observation 9 of the *Micrographia* deals with 'the colours observable in Muscovy glass, and other thin bodies'. This 'glass' is a mineral that is

---

* To give it its full title, *Micrographia; or Some Physiological Descriptions of Minute Bodies; Made by Magnifying Glasses, with Observations and Inquiries thereupon.*
† Hooke drew many of the illustrations himself; Wren helped with some of them.

'transparent to a great thickness', but is made up from many thin layers discernible under the microscope. Hooke was intrigued by the way this material converted white light into a rainbow pattern of colours, and discovered microscopic flaws in the layers of the material: 'with the *Microscope* I could perceive, that these Colours were ranged in rings that incompassed the white speck or flaw.' Newton, of course, is today remembered as the man who discovered that white light could be split into rainbow colours, and these rings are known, of course, as 'Newton's rings'. Hooke explained the phenomenon as a result of the combination (we would now say interference) of light reflected from the upper and lower surfaces of the thin layers, and described how the effect was only produced if the layers were thinner than a critical thickness; his explanation was based on the idea that light is a form of wave, in his words 'a very *short vibrating motion*', but incorrectly suggested that red and blue are the primary colours from which others are derived by 'dilutions'.

Even here, though, Hooke's reasoning was sound, given the state of knowledge at the time, and based on an experiment that clearly intrigued the young Isaac Newton. Hooke allowed a narrow beam of sunlight to enter the top of a conical flask filled with water, striking the surface of the water at an angle. He saw how the beam of light was spread out as it entered the water, producing a band of colour with red (he called it scarlet) on one side and blue on the other, with other fainter colours in between. It was this that led him to infer that white light is a mixture of colours (which is correct) and that red and blue are the primary colours, which are mixed together in different amounts to produce different colours (which was wrong, but not stupid). This experiment, described in Observation 9, is what pointed Newton towards his experiments with prisms, for which he is credited for the discovery that white light is a mixture of colours.

But the breadth of Hooke's interests and the depth of his theorising (the things that worried the Council of the Royal) can be seen in his summing up at the end of the Observation:

I think these I have newly given are capable of explicating all the *Phenomena* of colours, not only of those appearing in the *Prisme*, Water-drop or Rainbow, and in *laminated* or plated bodies, whether in thick or thin, whether transparent, or seemingly opacous.

The whole Observation amounts to what we would now call a scientific paper, and as 'Espinasse points out it is 'a progression of precise observation, masterly analysis and induction, and speculation'.

In Observation 58, one of the three astronomical observations, Hooke returns to optics to discuss the phenomenon of refraction, starting out from the by then well-known telescopic observation that 'the Sun and Moon neer the Horizon, are disfigur'd (losing that exactly-smooth terminating circular limb, which they are observ'd to have when situated near the Zenith)'. After discussing several other phenomena, notably 'that both fix'd Stars and Planets, the neerer they appear to the Horizon, the more red and dull they look, and the more they are observ'd to twinkle', he concludes:

First, that a *medium*, whose parts are unequally *dense*, and mov'd by various motions and transpositions as to one another, will produce all these visible effects upon the Rays of light, without any other *coefficient* cause.

Secondly, that there is in the Air or *Atmosphere*, such a variety in the constituent parts of it, both as to their *density* and *rarity*, and as to their divers mutations and positions one to another.

By *Density* and *Rarity*, I understand a property of a transparent body that does either more or less refract a Ray of light.

And

The redness of the Sun, Moon and Stars, will be found to be caused by the inflection of the rays within the

    Atmosphere . . . it is not merely the colour of the Air inter-
    pos'd.

In other words, the colour is inherent in the original white light
and is not some kind of pollution, or corruption, caused by the
passage of light through the intervening medium – another
discovery later attributed to Newton.

    The second great insight in *Micrographia* comes in Observation
16, where Hooke presents his ideas on combustion. The microscopic justification for including these ideas comes from his studies
of charcoal and burnt vegetables, but the experiments from which
his most impressive insights are drawn do not really involve the
microscope at all. These included his observations of the way
flames went out when a lit candle was shut in a sealed chamber,
how small animals collapsed and died after a certain time in such
a chamber, the gruesome vivisection of a dog, and the experiments
with candles and living things involving the air pump. Having
already, in Observation 9, asserted that heat is 'a motion of the
internal parts' of a substance (also mentioned in Observations 7
and 8), he now draws a clear distinction between heat and combustion. 'This *Hypothesis*,' he says, 'I have endeavoured to raise from
an Infinite of Observations and Experiments, the process of which
would be much too long to be here inserted.' But as he tells us,
the idea 'has not, that I know of, been publish'd or hinted, nay,
not so much as thought of, by any.' He was right.

    One of the key series of experiments that he hints at here was
carried out as demonstrations at the Royal in January and
February 1665. In a beautiful example of the scientific method at
work, he showed first that gunpowder would still burn in the
absence of air, and then that neither of two of the three ingredi-
ents of gunpowder, charcoal and sulphur, would burn on their
own in the absence of air. But each of them could be reignited
by adding the third ingredient, which he knew as saltpetre but
which we call potassium nitrate. As Hooke says in *Micrographia*,
it is clear from these experiments that combustion involves 'a
substance inherent, and mixt with the Air, that is like, if not the

very same, with that which is fixt in Salt-peter.' That substance is, of course, oxygen; the chemical formula for potasssium nitrate is KNO_3.

Hooke's idea is that something in the air is essential to combustion, which takes place when that something combines with something in the burning object. 'There is no such thing as an Element of Fire', he asserts, dismissing the idea that had held sway since the time of Ancient Greece. A flame 'is nothing else but a mixture of Air and volatile sulphureous parts of dissoluble or combustible bodies, which are acting upon each other whilst they ascend, that is, flame seems to be a mixture of Air, and the combustible volatile parts of any body'. Further, the component of air that is essential for combustion is also, Hooke tells us, essential for life. In Observation 22, almost as an aside, he mentions that there is a 'property in the Air which it loses in the Lungs, by being breath'd'. In being so close to the discovery of oxygen, Hooke was nearly a century and a half ahead of his time; right up until the end of the eighteenth century, the phlogiston theory of combustion (which said, flying in the face of experiments like those Hooke carried out with the air pump, that burning substances released phlogiston, rather than absorbing something from the air) held sway, and Hooke's ideas were forgotten. In 1803, chemist John Robison wrote:

> I do not know of a more unaccountable thing in the history
> of science, than the total oblivion of this theory of Dr. Hooke,
> so clearly expressed, and so likely to catch attention.

But it did catch the attention of one person, the serial plagiarist Isaac Newton. In an appendix to his book on optics, hurried into print immediately after Hooke's death (see postscript to Chapter Seven), Newton presented a suite of ideas about combustion that chemist Clara de Milt has described, with admirable academic restraint, as 'very, very much like those of Hooke'. As *Private Eye* might put it, could they by any chance be related?

The third great insight presented in *Micrographia* comes in

Observation 17: *Of Petrify'd wood, and other Petrify'd bodies.*
The petrified objects he refers to are what we now call fossils.
Before Hooke, it was widely thought that these were, in his words,
'Stones form'd by some extraordinary *Plastick virtue latent* in the
earth'. In other words, that these were just curious stones that
happened to resemble the forms of living things. But he dismissed
this notion, and stated unequivocally ('I cannot but think') that
they were 'the Shells of certain Shel-fishes, which, either by some
Deluge, Inundation, Earthquake, or some other such means, came
to be thrown to that place'. 'That place', he was well aware, was
high up in a mountain, or on the cliffs that he had walked as a
boy on the Isle of Wight. So how did such things as wood and
shells become petrified, or fossilised? Hooke's description of the
process could almost come from the pages of a modern textbook
of geology:

> this *petrify'd* Wood having lain in some place where it was
> well soak'd with *petrifying* water (that is, such water as is
> well *impregnated* with stony and earthy particles) did by
> degrees separate, either by straining and *filtration*, or perhaps,
> by *precipitation*, *cohesion* or *coagulation*, abundance of stony
> particles from the permeating water, which stony particles,
> being by means of the fluid *vehicle* convey'd, not onely into
> the Microscopical pores, and so perfectly stoping them up,
> but also into the pores or *interstitia*, which may, perhaps, be
> even in the texture or *Schematisme* of that part of the Wood,
> which, through the *Microscope*, appears most solid.

And as for shells, they must have been:

> fill'd with some kind of Mudd or Clay, or *petrifying* Water,
> or some other substance, which in tract of time has been
> settled together and hardened in those shelly moulds.

Hooke clearly understood two things: that there were geological processes that transformed once-living things into 'petrified'

rock, and that there were geological processes that transformed the structure of the Earth's crust. Implicit in this was the understanding that the timescales involved ('tract of time') were far greater than the 'official' chronology of a few thousand years derived from the Bible.

Hooke even begins to hint at the kind of investigations that would lead to the idea of evolution:

> It were therefore very desirable, that a good collection of such kind of Figur'd stones were collected; and as many particulars, circumstances, and informations collected with them as could be obtained, that from such a History of Observations well rang'd, examin'd and digested, the true original or production of all those kinds of stones might be perfectly and surely known.

Soon after the publication of *Micrographia*, a Dane, Niels Steensen (who used the Latinised version his name and is remembered as Steno), publicised very similar ideas, and suggested that different rock strata, containing fossils such as sharks' teeth, had been laid down under water, far from the present-day seas, at different times during Earth's history by a succession of floods. Coincidence? Hooke didn't think so. He had developed these ideas further in his Cutlerian Lectures which we discuss later. Henry Oldenburg, the Secretary of the Royal Society and someone who often rubbed Hooke up the wrong way, was in correspondence with scientists across Europe as part of his job. Steno published his ideas in 1669, in Latin. Oldenburg promptly made the Royal aware of the book, and arranged for it to be translated into English, which helped to ensure that Steno became remembered as the inventor, or discoverer, of these ideas. Hooke was not exactly pleased and tried unsuccessfully to get recognition that he at the very least had the idea first. When it was suggested that he had borrowed his ideas from Steno, rather than the other way around, he was moved to write a letter, read to a meeting of the Royal on 27 April 1687, in which he said:

I must now add in my own vindication that I did long since prove Steno had much of his treatise from my Lectures, which some time before that I had read [in Gresham College] which Lectures Mr Old. Borrowed and transcribed and by Divers circumstances I found he had transmitted the substance of if not the very Lectures themselves [to Steno]. And he did as good as own it, and upon my challenging him with it he did in two of his transactions publish that I had Read A great part of that Doctrine & hypothesis in my Lectures in Gresham Colledge Some time before Mr Steno had published his Booke.

There is no reason to doubt Hooke's version of affairs, and there is no doubt at all that his work preceded that of Steno, whether or not Steno got word of it via Oldenburg. Steno, by the way, never gave a clue one way or the other: he disappeared from the scientific scene after writing his book. He became a Catholic priest in 1675, and was ordained as a bishop in 1677, inflicting on his body such a harsh regime of fasting and self-denial that he died in 1686, at the age of forty-eight.

Hooke's more extensive ideas about earthquakes, Earth history and geology will be covered in Chapter Nine. Now, we still have a fourth great insight from *Micrographia* to discuss, although here we diverge from 'Espinasse's assessment of which of the ideas Hooke presented there were most significant. She picks out his discovery of the structures he named cells (after the rooms occurred by monks in a monastery) in thin slices of cork (Observation 18). As Hooke puts it, no 'Writer or Person' had 'made any mention of them before this'. But although the name was taken up and used by later biologists, it was in a slightly different context. The 'pores', as he also called them, that Hooke had found are not living cells, but non-living structures left over from the growth of the plant. The first person to see and study live cells under a microscope was Hooke's Dutch contemporary Antoni van Leeuwenhoek. In 1674 he described an algae, Spirogyra, and other organisms that moved of their own volition; he named them animalcules ('little animals'). In this area, Hooke's work was important, but not as important as the work of van Leeuwenhoek and others. In our estimation, his astronomical Observations were of far greater importance.

Hooke was a serious and highly respected astronomer. On 9 May 1664, using a twelve-foot-long refracting telescope, he had discovered the Great Red Spot of Jupiter, and used it to measure the rotation of the giant planet. Contemporary (and now more famous) astronomers such as the Italian Giovanni Cassini picked up on the discovery, and referred to the phenomenon as 'Hooke's Spot'. But it was observations of something much closer to home that led Hooke to important insights that appeared in Observation 60: *Of the Moon.* This is a short contribution that to a casual glance looks like a mere filler. That couldn't be more wrong.

Observation 60 provides a nice example of the scientific mind – Hooke's scientific mind – at work: making observations, devising hypotheses, testing them by experiment and further observation, and drawing general conclusions from specific cases. Remember that this was less than sixty years after Galileo, with the aid of one of the first astronomical telescopes, discovered that the Moon is not a perfect sphere but pockmarked with craters and scarred by mountain ranges. Hooke was intrigued by the nature of these craters, and puzzled over their origin. He described them as 'almost like a dish, some bigger, some less, some shallower, some deeper, that is, they seem to be a hollow *Hemisphere*, incompassed with a round rising bank, as if the substance in the middle had been digg'd up, and thrown on either side'. Which establishes, as if we did not already know, that he was a good and accurate observer.

How could such craters be formed? Hooke came up with two hypotheses and set out to test them. The first was that the craters were caused by impacts. To test this, Hooke made a mixture of water and pipe-clay, 'into which, if I let fall any heavy body, as a Bullet, it would throw up the mixture round the place, which for a while would make a representation, not unlike these of the Moon.' So incoming objects (bodies) would do the trick. But

Hooke found it 'difficult to imagine whence those bodies should come', so he turned to his other idea. In this experiment he heated a pot of alabaster to the boiling point, and then, while it was still bubbling, took it off the fire and allowed it to set. Then 'the whole surface, especially that where some of the last Bubbles have risen, will appear all over covered with small pits, exactly shaped like these of the Moon, and by holding a lighted Candle in a large dark Room, in divers positions to this surface, you may exactly represent all the *Phenomena* of these pits in the Moon, according as they are more or less enlightened by the Sun'.

So Hooke plumped for volcanic activity as an explanation of lunar cratering, rather than impacts. This was a perfectly reasonable conclusion to draw at the time, and for the next four hundred years volcanic activity remained a viable explanation for lunar cratering. The idea was only finally laid to rest, in favour of the impact hypothesis, when astronauts visited the Moon and its geology could be studied first hand. We now know that the craters were indeed made by impacts, in which 'the substance in the middle had been digg'd up, and thrown on either side'. But if it was thrown up, either by impacts or by volcanic activity, something must have pulled it back down on to the surface of the Moon to make the circular ramparts surrounding the craters. That something, Hooke reasoned, must have been gravity – the Moon's own gravitational pull.

Developing his idea, Hooke said that it 'is not improbable, but that the substance of the Moon may be very much like that of the Earth' (which would have amounted to heresy a few decades earlier). And then he goes beyond Galileo, who noticed the imperfection of the Moon, to draw attention to the remarkable roundness of the Moon in spite of the small irregularities we see on its surface. The Moon, he points out:

we may perceive very plainly by the *Telescope,* to be (bating the small inequality of the Hills and Vales in it, which are all of them likewise shaped, or levelled, as it were, to answer to the center of the Moons body) perfectly of a Spherical figure, that is, all the parts are so rang'd (bating the comparatively small ruggedness of the Hills and Dales) that the outmost bounds of them are equally distant from the Center of the Moon, and consequently, it is exceedingly probable also, that they are equidistant from the Center of gravitation; and indeed, the figure of the superficial parts of the Moon are so exactly shap'd, according as they should bye, supposing it had a gravitating principle as the Earth has.

This is mind-blowing stuff. At a time when other people talked about vortices and whirlpools being responsible for the shape of the planets and their orbits, and Isaac Newton was an unknown student who would soon be eagerly devouring Hooke's book,* Hooke is suggesting the universal principle of gravitation (he can hardly have failed to notice that Jupiter and the other planets are also round!), that all objects possess this property, which makes the moons and planets round and (although he discusses this elsewhere) holds them in their orbits around the Sun. The very last paragraph of *Micrographia* begins with the words: 'To conclude, therefore, it being very probable, that the Moon has a principle of gravitation . . . whereby it is not only shap'd round, but does firmly contain and hold all its parts united, though many of them seem as loose as the sand on the Earth'.

We emphasise that the idea of *universal* gravity is of key importance. This is the beginning of an understanding that the laws of physics which operate in the Universe at large – in the Heavens – are the same as the laws that apply here on Earth. That idea is often traced back to Newton; it should be traced back to Hooke. It's a long way from the study of the point of a needle!

Any of these four ideas, or indeed his ideas about planetary orbits and gravity, which we have already discussed, should have ensured Hooke's status as one of the greatest scientists of all time. And remember that there were dozens of lesser 'observations' in

---

* Newton received his BA in January 1665, the same month that *Micrographia* was published.

Micrographia, including some of the first observations of the tiny
creatures that live in water and other liquids – the Fellows were
particularly intrigued by the discovery of the creatures we call
nematodes, but were referred to then as 'eels', living in vinegar
(Observation 57). Perhaps Hooke would have been suitably recognised by posterity if he had been able to develop his ideas more
fully, which he clearly intended to do. In his book, he says (especially in reference to his ideas about combustion, but undoubtedly
with broader relevance):

> In this place I have only time to hint at an *Hypothesis*, which,
> if God permit me life and opportunity, I may elsewhere
> prosecute, improve and publish.

But Hooke's opportunities to 'prosecute, improve and publish'
his revolutionary ideas were almost immediately restricted by
plague, the Great Fire of London, and a change of career that
occurred in the aftermath of the fire. While he was otherwise
engaged, at least some of those ideas were taken up and developed
by Isaac Newton, who had an early copy of *Micrographia* which
he read and annotated extensively (the copy still exists), having
ample opportunity to study it while he was away from Cambridge
during the plague year of 1665 when he was twenty-two. He was
particularly inspired at that time by Hooke's ideas about light
and colour, developed in Observation 10. Newton's variation on
this theme would soon come to the attention of the Royal and
lead to a lifelong bitterness between Hooke and Newton. But
Hooke would have ten more years – the happiest years of his life
– before that controversy reared its head.

CHAPTER THREE

MONUMENTAL ACHIEVEMENTS

In the spring of 1665, following the publication of *Micrographia*, Hooke's prestige was higher than ever, and he was, at the age of twenty-nine, at the peak of his abilities as a scientist and 'mechanick'. But his immediate plans, and those of the Royal Society, were disrupted by the severe outbreak of plague that affected London as the weather got warmer. Plague was far from being unknown, and there had been lesser outbreaks from time to time, but this occasion was different. Cases had been reported in Holland in the spring of 1664, resulting in ships from the Netherlands being quarantined in the Thames; in February 1665, war broke out between England and the Dutch, so all trade ceased. In any case, the cold winter of 1664–1665 had slowed the spread of the disease, but the death toll began to rise in March 1665.

At the time, the Lord Mayor was Sir John Lawrence, who that same month had been instrumental in rectifying the irregularities involving Hooke's election as Professor of Geometry, and he stayed in the City to keep control throughout the months that followed;

Henry Oldenburg, the Secretary of the Royal, also stayed at his post, acting as a conduit (the seventeenth-century equivalent of the Internet) for the flow of information between scientists in Europe (including Holland, in spite of the war) and England.* Samuel Pepys, of course, was another who stayed. But those who could leave the city did so. The King and his Court moved to Oxford, followed by many of the Fellows of the Royal, including Boyle. Queen Henrietta went to Paris, accompanied by a large party including Christopher Wren, whom the King had instructed to study the buildings programme of Louis XIV. But after the Royal suspended its Wednesday meetings following 28 June, its three leading experimenters – Hooke, Sir William Petty and Dr John Wilkins – moved (with a lot of experimental apparatus, and an assistant, known as an 'operator') to Durdans, an estate near Epsom in Surrey, owned by George, Lord Berkeley, himself an FRS. There, they carried out a full programme of experiments on behalf of the Royal, and many others of their own, particularly Hooke's, devising.

The work that was of greatest interest to the Royal is of no interest to us. Always eager to demonstrate the practical value of their research, the Society urged the experimenters to devise improved, lightweight carriages that would be both fast and comfortable for their passengers. Their designs were successful, but did nothing to advance the progress of science. A related development, which Hooke worked at on and off for years, was a 'waywiser' to measure how far a coach had travelled, based on counting the rotations of a wheel attached to the vehicle. Another potentially practical (and lucrative) project commissioned by the Royal was a series of experiments aimed at improving marine timekeepers. But as we have mentioned, these ultimately failed to solve the longitude problem, partly because the pressure of other work prevented Hooke from concentrating on clocks, and partly because of the disagreements about patent rights.

---

* Michael Cooper has suggested, in the light of what we know of Oldenburg's character, that this was less out of a sense of duty than because Oldenburg loved sitting like a spider at the centre of the scientific web, privy to all the gossip and intrigues (which, as we shall see, he was not averse to stirring), and feared being replaced.

Boyle visited Durdans in July 1665, and carried out some experiments with Hooke (that is, as usual Boyle and Hooke together devised the experiments, then Hooke and the operator did the work). These helped him to finish his book *Hydrostatical Paradoxes*, but he left early in August, and Petty went on to Salisbury a day or so later. Hooke had increasing freedom to carry out his own experiments. These included astronomical observations and the development of new (or improved) astronomical instruments, building on his work with Wren. One of the motivations for this work was again practical: if measurement could be made accurate enough then it would be possible in principle to determine longitude by timing the exact moment when the edge of the Moon passed in front of (occulted) stars whose positions on the sky were already known.*

But for us the most interesting work that Hooke carried out while at Durdans involved his continuing investigation of gravity, making use of two deep wells on nearby Banstead Downs. First, he investigated claims made by Henry Power, three years earlier, that the weight of an object is less underground than at the surface of the Earth. But as Hooke wrote to Boyle on 15 August 1665:

> I have made trial since I came hither, by weighing in the manner, as Dr. Power pretends to have done, a brass weight both at the top, and let down to the bottom of a well about eighty foot deep, but contrary to what the doctor affirms, I find not the least part of a grain difference in a weight of half a pound between the top and bottom. And I desire to try that and several other experiments in a well of threescore fathom deep, without any water in it, which is very hard by us.

The other well turned out to be blocked at a depth of 315 feet, and the experiments tried there also failed to show any change in the weights. But this was not unexpected: Hooke knew that

---

* This idea was important in the development of Edmond Halley's career; see Chapter Eight.

much more sensitive methods would be needed to measure any changes, and when he presented his findings to the Royal in March 1666, he offered some suggestions for the kind of instruments that would be required (the designs were sound, but beyond the technology of the time). But the key passage of the paper *On Gravity* that he presented to the Royal that month shows his understanding of the nature of gravity:

> A body at a considerable depth, below the surface of the earth, should lose somewhat of its gravitation, or endeavour downwards, by the attraction of the parts of the earth placed above it.

This is another version of Hooke's realisation of the *universal* nature of gravity. It was not a mystic force, pulling things only to the centre of the Earth (or, indeed, the Moon), but a property of *all* matter, with the material above the object pulling upwards just as the material below the object pulled downwards. Twenty years later, Hooke's priority in understanding this would play a key part in his dispute with Newton (see Chapter Seven).

In the middle of these experiments, Hooke visited the Isle of Wight in the autumn of 1665. His mother had died in June, but at that time travel to the island was restricted in the hope of preventing the spread of plague. It was only in October that Hooke was able to visit his childhood home to settle family matters. It was at this time that he made a more careful investigation of the fossils that had intrigued him as a boy, making notes and sketches as he walked around the south-west corner of the island. This led to the first of his 'Discourses on Earthquakes', presented in lectures starting in 1667, but only published after his death; we shall save these to describe together later (Chapter Nine). Hooke was back at Durdans by January 1666, and returned to Gresham College the following month, just before Wren returned from Paris and the King from Oxford. But the City had just six months to get back to normal life before a disaster in some ways worse than the plague struck.

During these months Hooke was active and prolific. As well as his discovery of the Great Red Spot of Jupiter, which had been made earlier but was published in March 1666, he measured the rotation of Mars, continued his work on timekeepers, carried out attempts at blood transfusions with dogs (some more successful than others) and gave lectures. Pepys mentions a 'very pretty' lecture on the trade of felt-making, a reminder to us of the practical side of Hooke's work. But we want to focus on one achievement in particular from that summer, yet another invention that should have made Hooke even more famous than he was but was overlooked; this time it was neglected not least because of the fire which broke out on Sunday 2 September 1666. At the meeting of the Royal scheduled for 12 September (which was, of course, abandoned in the aftermath of the fire) Hooke had been intending to describe what was then called a reflecting quadrant, but developed into the instrument which we now know as a sextant.

This was another astronomical instrument that doubled as an aid to navigation, measuring not longitude (the distance east or west of the home port) but latitude (the distance north or south of the equator). Latitude can be determined by measuring the height of the Sun above the horizon at noon; in astronomical work the same kind of instrument can be used to measure the height of a star above the horizon, or the angular distance between two astronomical objects. Prior to Hooke's invention, this involved looking through a telescope or open sight at the horizon, holding the instrument steady (easy for astronomers on solid Earth, less easy for mariners on the deck of a heaving ship), and moving a second telescope (or sight) on an arm hinged to the first one on to the target star (or Sun), then measuring the angle between the two jointed arms. Hooke's ingenious idea was to attach a small mirror to the moveable arm of the instrument, which reflected an image of the object being observed (the target star, or the Sun) on to a second mirror and along the first (indeed, now the only) sighting arm, while this was pointed at the horizon. The operator could then see both the horizon and an image of the target along the same sight, and the angle of the moveable arm could be adjusted until the image of the target seemed to be sitting on the horizon. The angle when this occurred could be read off from a curved, graduated scale on the instrument.*

Hooke's main interest in developing the instrument was astronomical – he told the Society, in the run-up to the intended lecture, that he had a method to determine accurately the angular distances of stars relative to the Moon, which had a direct bearing on the longitude problem. But had he not been distracted by other work, and been able to develop the idea, it would surely soon have been applied to navigation. Hooke's instrument was so completely forgotten that in 1691 Edmond Halley reinvented it, but quickly withdrew his claim for priority when it was pointed out that his friend Hooke had come up with the idea twenty-five years earlier. And in 1699, Isaac Newton claimed to have invented such a device, which roused the by then ailing Hooke to drag himself along to the next meeting of the Royal to remind them of his priority. The idea only really took off, however, after 1731, when John Hadley demonstrated his own version of the sextant to the Royal Society.

In September 1666, the City of London was, as far as fire was concerned, an accident waiting to happen. Many of the lanes were only five or six feet wide, and the upper storeys of the buildings hung over the lanes so that they almost touched, perhaps with a foot of space through which a strip of sky might be visible. The houses themselves were timber-framed, with a lathe and plaster infill. And everything was tinder dry, following a long, hot summer. In addition, on 2 September, when fire broke out in a bakehouse in Pudding Lane, there was a strong north-easterly wind, which fanned the flames and swept them westward and down to the Thames. The conflagration raged for three days before a slackening wind enabled the firefighters to get on top of it, and by Thursday most of the City was a smouldering ruin. The Royal Exchange, source of Sir Thomas Gresham's wealth,

---

* The curved scale covered one-quarter of a circle, ninety degrees, which gave the quadrant its name. The more compact version developed later covered sixty degrees, one-sixth of a circle, hence the name sextant.

was gone, but the fire had stopped about 200 yards from Gresham College, in the north-eastern (upwind) quarter of the City. Only about 75 acres out of just over 400 acres within the City walls survived, while a further 63 acres to the west, covering Holborn and Fleet Street, had also gone. The old St Paul's Cathedral was lost, together with 90 churches, 52 Livery Company halls and more than 13,000 houses.* An estimated 100,000 people lived in the City at the time, and three-quarters of these were now forced to camp outside in the open spaces beyond the walls, but fortunately there was only relatively minor loss of life, perhaps a few hundred people.

The City was governed by a Court of Aldermen, or corporation, under the Lord Mayor; having lost their Guildhall home, the Court moved into Gresham College, meeting there for the first time on Thursday 6 September, and forcing all the professors and their tenants (legal and illegal) out, except for the household of one man – Robert Hooke, who had good City connections and was seen as a valuable person to have on hand for advice and help. The Royal was able to find accommodation in Arundel House on the Strand, and continued to function – which meant that Hooke and his operator had to trek about a mile and a half across the ruins and up Fleet Street with the equipment for the demonstrations he was still required to perform. The astonishing thing is that he did continue to carry out a full programme for the Royal, and give his Gresham and Cutlerian lectures, even though he now undertook a task that to anyone else would have been a full-time occupation: supervising the rebuilding of London.

While the fires were still smouldering, several people presented plans for the rebuilding of the city to the King. The City† had moved with impressive speed to restore order, issuing instructions for clearing the rubble, setting up temporary markets, and so on. They recognised the need for swift action on a formal rebuilding

---

* There is some uncertainty about most of these numbers; we follow Michael Cooper's assessment.

† We use the term 'City' for the Court of Aldermen, to avoid confusion with the Royal Court.

programme to prevent unregulated reconstruction higgledy-piggledy across the city, and on 13 September in a wide-ranging proclamation in support of their actions, Charles II instructed the Lord Mayor and City to:

> cause an exact survey to be made and taken of the whole ruins occasioned by the late lamentable fire, to the end that it may appear to whom all of the houses and around did in truth belong, what term the several occupiers were possessed of, and what rents, and to whom, either corporations, companies, or single persons, the reversion and inheritance appertained: that so provision may be made, that though every man must not be suffered to erect what buildings and where he pleases, he shall not in any degree be debarred from receiving the reasonable benefit of what ought to accrue to him . . . we shall cause a plot or model to be made for the whole building through those ruined places: which being well examined by all those persons who have most concernment as well as experience, we make no question but all men will be pleased with it, and very willingly conform to those orders and rules which shall be agreed for the pursuing thereof.

The initial hope of the King and the City was to rebuild London on a grand scale as a completely new city, but they realised very quickly that speed was of the essence, and that the rebuilding programme must start as soon as possible (which meant in the spring of 1667, when the winter was over) or haphazard illegal construction would begin, and the homeless citizenry would become restless. It was with this in mind that the plans submitted while the ground was still hot were, though widely admired, rejected. First off the mark was Christopher Wren, who presented his idea for a new city to the King on 11 September; he was quickly followed by John Evelyn, who presented his own proposal to Charles two days later. These direct approaches to Charles upset Oldenburg, who felt that an opportunity to promote the

Royal Society had been missed. Both Wren and Evelyn were Fellows, and if the plans had gone first to the Royal and then on to the King in the name of the Royal it would have benefited the Society. Hooke was more diplomatic. His plan, probably endorsed by Sir John Lawrence, was shown first to the City and then to the Royal, at a meeting held on 19 September. Only then was it presented to the King. Hooke's proposal, we are told by Richard Waller, was for a rectangular layout of streets, a grid like the layout of many modern American cities, but the original plan has been lost. In any case, all these ideas (and a couple of others) came to naught. Because of the need for speed, it was decided to rebuild the city essentially along the old lines, literally building on the old foundations, but with some streets being widened and proper provision being made for facilities such as markets. The joint responsibility for getting this done was shared by the Privy Council (on behalf of the King) and the Court of Aldermen, representing the City. On 4 October 1666, a meeting of the Court considered a report from the Privy Council, and made the declaration that since:

for the better and more expedition of this work [the King] hath pleased to appoint Dr Wren Mr May and Mr Pratt to joyne with such Surveyors & Artificers as should be appointed by the City to take an Exact & speedy survey of all Streetes Lanes Aleys houses & places destroyed by the late dismall Fire That every particular Interest may be ascertained & provided for & the better Judgment made of the whole Affaire This Court doth therefore Order that Mr Hooke Reader of the Mathematicks in Gresham Collidge Mr Mills and Mr Edward Jermyn do joyne with the said Dr Wren Mr May & Mr Pratt in taking the said Surveigh . . .

There was clearly a delicate balancing act going on here. The three Royal appointees, known as 'the King's Commissioners',*

---

* Formally, 'His Majesties Commissioners for Rebuilding'.

had all been advising Charles, before the fire, on plans to repair the old St Paul's Cathedral, which was now beyond repair and would have to be replaced by a new cathedral. The three City appointees, known as 'the City Surveyors',* were chosen to match the number of King's Commissioners, and perhaps with an eye on matching them intellectually, or academically, as well. Peter Mills was an obvious appointment; he was already the City Surveyor, and one of the people who had drawn up a plan for the rebuilding. Edward Jermyn (or Jerman) turned down the appointment (probably because he had a busy and lucrative private architectural practice); the fact that he was not replaced suggests that, indeed, the City had only appointed three men in the first place because the King had three. Hooke was a less obvious choice, in that he had only limited experience of building work, but he was known as a very able man who got things done – the only Gresham Professor who had been allowed to stay in residence, precisely because he might come in useful. He was also a friend and academic partner of Wren, his intellectual equal; it seems likely that these factors encouraged the City to appoint him, partly as a counterweight to Wren who could argue the City's case if necessary, partly that their friendship might make such arguments less likely. Michael Cooper, the leading authority on Hooke's work as City surveyor, describes the appointment as 'remarkable', highlighting that:

> A man with, as yet, no formal connection with the City except his appointment at Gresham College was trusted effectively to determine on the City's behalf all technical aspects relating to the rebuilding of the city.†

This speaks volumes for Hooke's standing in the community, which rose even higher as he carried out the commission. Whatever the City's motives, the choice was inspired, and created a

---

* Formally, 'Surveyors of New Buildings'.
† 'Robert Hooke's work as surveyor for the City of London in the aftermath of the Great Fire'. Part one. *Notes and Records of the Royal Society*, volume 51, pp 161–174, 1997.

partnership that was spectacularly successful. Oldenburg, no particular friend of Hooke but an astute observer, commented in a letter to Boyle with news of the surveying and rebuilding plans that it was 'by the care and management of Dr Wren and M. Hook' that it would be carried out.

As Jardine comments:

The arrangement was an extremely fortunate one for the Corporation of London: Hooke answered directly to them, and was tireless in his work facilitating swift rebuilding by individual householders. Wren had the ear of the King.

Hooke and Mills began work immediately but were not formally sworn in as City Surveyors until 14 March 1667, when their pay was initially set at £100 a year (soon increased to £150), and arrears back to September 1666 were paid. Throughout his time as City Surveyor Hooke was paid properly and on time; this was the beginning of an accumulation of considerable wealth and the end of any financial worries he might have had. There were also other fees associated with the job. During his full-time work on the rebuilding,* up to the end of 1673, Hooke received a total of £1,062, with some lesser payments for odd bits of work carried out over the following ten years. In June 1667, John Oliver, an experienced surveyor, joined the team to assist Mills, who was increasingly infirm and died in 1670, when Oliver replaced him. So Hooke was very much the senior surveyor on the City side.

In the autumn and winter of 1666–67, before the rebuilding began (and before he was paid!), Hooke tackled a variety of tasks with his usual enthusiasm and efficiency. And he did this while fully committed to his other life. Throughout the period of his work on the rebuilding of London, Hooke mostly carried out his City work in the mornings and his work for the Royal, which we discuss later, in the afternoons; he slept very little, being a chronic insomniac. He was asked to experiment with brickmaking

---

* 'Full-time' apart from his scientific work and lecturing, that is.

and report which were the best bricks to use; he worked with Mills to specify building regulations governing such things as the height of the new houses, the materials to be used in their construction, and removing the physical irregularities that had existed before the fire. In the old city, upper-storey rooms for one house might overlap the lower-floor rooms of another, and be overlapped in their turn still higher up; staircases were uneven; chimneys might have no clear ownership, and so on. In the new city, walls would be vertical, with no such irregularities. The regulations also provided for the arbitration process to be used when the inevitable disputes arose.

Another commission did not come from the City, but could not be ignored, given Hooke's status as a Gresham Professor. The Gresham Trustees asked for advice on the cost and design for a rebuilt Royal Exchange. In November 1666 Hooke came up with a detailed proposal for a beautiful building, with an estimated cost of under £5,000, saving money by re-using some of the old materials. The Trustees rejected the proposal and opted for a lavish new building, designed by Jermyn, which cost so much (roughly £60,000) that servicing the debt caused a decline in the once-great Mercers' Company. The new Exchange was, however, fully open by 1672, allowing Gresham College to get back to normal. But from our point of view, what is interesting is that the skill displayed by Hooke in carrying out this commission shows that he already had a thorough grasp of architecture, the practical aspects of building, and the strength of materials.

The first practical steps towards the rebuilding of London started on 27 March 1667, financed by a tax on coal. As well as the practical need to wait for spring, nothing could be done on the ground before the appropriate legislation had been issued by Parliament; the relevant Act had received the Royal Assent on 8 February. Then, before anything could be built, the survey had to be carried out. This was a major undertaking. The surveyors – Hooke and Mills – started work with Fleet Street, which was the main route from the City to Westminster, and would be widened to about fifty feet. The surveying involved seven labourers to clear the rubble before the surveying work could take place, and six carpenters to cut the wooden stakes used to mark the boundaries. After the initial work on Fleet Street, the surveyors worked independently of one another, with Mills, as we have mentioned, soon being replaced by Oliver. Most of the work was completed by the end of 1671, and the largest individual portion was carried out by Hooke. In all, 8,394 foundations were surveyed and staked out for rebuilding; of these, 1,582 are recorded as being Hooke's responsibility, and just over 2,800 are credited to one or other of Mills and Oliver. That leaves about 4,000 unattributed surveys. One reason that many of those are unattributed is that Hooke, unlike the other surveyors, did not hand in his surveying notebooks to the authorities, and they have disappeared; he probably retained them because he used them for making scientific notes, which he wanted to keep, as well as recording his surveys. But from other evidence (including Hooke's diary, more of which shortly) it has been estimated that at least 1,400 of these 'belong' to Hooke, giving him a total of some 3,000 surveys. Cooper has estimated that for rather more than five years this work occupied Hooke for three hours a day, six days a week.

But that was not the full extent of his labours on the ground. When disputes arose during the course of building, at least one of the surveyors had to make a site visit to view the situation and resolve the difficulty, writing a report and issuing a certificate to the successful party in the dispute. The householder (or tenant) had to pay a fee for these views and certificates, part of which went to the surveyor. This kind of work did not always involve disputes between neighbours: Hooke also received fees from individuals for staking out and certifying the foundations of properties prior to building. Hooke always wrote the reports if he was involved in the views, even if another surveyor also took part, and it seems that he also collected the fees on behalf of the City, and passed the appropriate share on to the other surveyor if they had been involved (among the characteristics that made Hooke so valuable to the City were his scrupulous honesty and accurate record-keeping). This work continued intermittently long

after the fire. In a diary entry dated 6 May 1693, Hooke refers
to a meeting with Oliver at Jonathan's coffee house:

> I paid J. Oliv. for yesterdays View 10 [shillings]. Viewed it
> again with J.O. I drew Report at Jonathan's, we both signed
> it.

So Hooke, more than anyone, was responsible for ensuring that
building regulations were obeyed and that the new city took on
the appearance that it did. Hooke made at lest 500 views of new
buildings himself, mostly between 1668 and 1674, earning at least
£1,600 in fees for the reports and payments for certificates issued
to property owners and tenants allowing them to begin work, or
to claim compensation for land lost to improvements such as road
widening. By the end of 1671, more than 7,000 houses had been
built, and the city was getting back to normal, but now with brick
buildings, wider roads, new marketplaces, and the gradients of
the steeper hills eased. If Hooke had made no other contribution
to the rebuilding, his major contribution to this would be a stag-
gering achievement worthy of far more recognition than it has
generally received. But he did do more, in a collaboration with
Wren, who has sometimes overshadowed his contribution to city
planning almost as effectively as Isaac Newton overshadowed his
contribution to science, but without Newton's malicious intent.

There are four aspects of Hooke's work with Wren that are
particularly impressive, ranging from the practical to the sublime.*
The practical project involved restoring the Fleet Stream from its
state as a refuse-filled ditch to a navigable river or canal; the fact
that for reasons beyond the control of the architects it later fell
back into decay does not detract from their achievement.

By 1666 the Fleet was accurately described by the common
name of the Fleet Ditch, although it had once been a channel that
could be navigated at high tide from the Thames to Holborn

---

* We have picked out those examples, but they are not exhaustive; for a full account of
Hooke's architectural work, see Cooper.

Bridge. If the city was being rebuilt anyway, surely here was an opportunity to restore the river to its former glory, canalising it and providing broad quays for unloading goods. The project, which on the City side was largely promoted by Sir John Lawrence, was assigned to Hooke and Wren, who made detailed proposals, which were approved by the City in March 1671. The project was beset by many problems, not least the fact that although it was easy to clear rubbish from the ditch and start digging, it soon filled up again with detritus carried by runoff from rainfall down the steep sides of the little valley in which the stream lay. In addition, since the remit of the City only extended as far as Holborn Bridge, there was nothing to stop people further upstream continuing to dump their household refuse, industrial waste, slaughterhouse offal and other unsavoury items into the water to be carried downstream, and although gratings were built to hold back this flow, they often broke under the pressure.

In spite of all the difficulties, the project was completed by the autumn of 1674, at a cost of £51,307 6s 2d. It was intended to recoup the cost of maintaining the canal by charging tolls for the use of the quays by barges, but it turned out to be quicker (and cheaper!) for barges to be unloaded at the river and goods to be transported inland by carts. Indeed, the broad quays lining the Fleet Canal (as it now was) became used as highways for wheeled traffic heading north and south. The canal fell into disrepair and once again the Fleet became filled with silt and rubbish; it was eventually (in 1769) covered by a road, but the Fleet still flows underground beneath Farringdon Street and New Bridge Street.

Hooke was also involved in other civic works, not least laying pipes to bring fresh water to the population, and sewers to remove waste. But the second achievement of the Wren–Hooke partnership that we want to mention is a well-known one, but in which Hooke's role is usually greatly underestimated. We refer, of course, to the rebuilding of London's churches.

The authorities responsible for the rebuilding of parish churches, appointed by Parliament, were the Archbishop of Canterbury, the Bishop of London and the Lord Mayor; they became known as the Commissioners for Churches, although this was not an official title. In their turn, in 1670 they put Wren in charge of the design and construction of the churches, with Hooke and Edward Woodroofe named as his assistants. Woodroofe was already working with Wren as his assistant on the St Paul's project, and did little work on the churches; he died in 1675 and was replaced by John Oliver, who was the kind of assistant that the term usually implies. The working relationship between Wren and Hooke was, however, much closer and more equal than the term 'assistant' usually implies. The notion that Hooke was not involved in the design of the buildings, but merely supervised the construction following Wren's orders, stems from a book, *Paternalia*, written by Wren's son and published in 1750, which plays Wren's role up and Hooke's role down. But modern (unbiased) authorities have a quite different view: in modern terms, it would be better to regard Hooke as a junior partner in Wren's architectural firm, although not much the junior.

Although most of the parish churches were built of stone, many suffered severely in the fire, with burnt-out interiors, collapsed roofs, damaged steeples and unstable walls. Out of 110 churches, twenty survived the fire, thirty-five were demolished and not rebuilt, and fifty-five were pulled down and rebuilt. But this programme of rebuilding did not begin immediately after the fire, partly because of the cost and partly because of the pressing need to rehouse the people of the City (in the interim, they worshipped in temporary wooden chapels). The money was raised following an Act of Parliament in 1670, imposing a further tax on coal for this specific project. The work continued until 1695, alongside Wren's work on other projects, including St Paul's, and it is impossible that he could have been responsible for all of the design work. A handful of churches, notably St Benet Paul's Wharf (probably the best surviving example in London if you want to see what a Hooke church looks like), St Edmund the King, St Martin within Ludgate and St Michael's Crooked Lane, are now confidently identified as largely or solely Hooke's work, and many others were likely to have been chiefly his or a result of a collaboration with Wren. On the other hand, the documentary evidence cannot unambiguously identify Wren as the sole or chief 'author' of more than half a dozen of the churches. Most of the churches have little or no documentation and there is no way to tell for certain from other evidence who worked on them, but Hooke seems to have been influenced by Dutch design, which has helped in the attribution of some of the rest. Partly on this basis, in his book *The City Churches of Sir Christopher Wren* Paul Jeffery argues that Hooke and Wren divided the City up between them, roughly fifty–fifty, with Hooke attending to churches in the north and east (near to his base at Gresham College) and Wren concentrating on the south and west. This seemingly plausible division of labour into rough, if not exact, equality has been questioned, but, coming as it does from an author writing in praise of Wren rather than Hooke, it has to be taken seriously. It also matches the equal division of responsibility between Hooke and Wren in their collaborations on astronomy and other scientific work.

The most compelling evidence, though, is financial. Between 1671 and 1693 Hooke was paid a total of £2,820 from the fund for 'Officers and Servants Emploid in Building Parochiall Churches'. This is more than the *combined* income he received as Gresham Professor and as City Surveyor, and far more than he would have earned merely by supervising the work on the ground. As Stephen Inwood has pointed out, we can only conclude 'that his career as a City church builder was as important as his career as a scientist, a lecturer, or a City Surveyor.'

Hooke certainly had a high and justified contemporary reputation as an architect, and received several significant private commissions. Among the most important of these were: the Bethlehem Royal Hospital (known as 'Bedlam') for the mentally ill, in Moorfields; Aske's Hospital, in Hoxton; Merchant Taylor's School, in Suffolk Lane; the College (later Royal College) of Physicians, in Warwick Lane; Bridewell Hospital; and (by no means least) Montagu House, in Bloomsbury. None of these survives. Outside London, some of Hooke's buildings do still stand: the almshouses commissioned by Seth Ward in Huntingdon; Ragley Hall, in Warwickshire; Ramsbury Manor, in Wiltshire; Sheffield Place, in Essex; and our favourite, the little church he built for Richard Busby at Willen, in Buckinghamshire. Even this is not an exhaustive list, and Cooper has estimated that in addition to the salary he received for his work on the city churches Hooke received about £2,000 for his private architectural and building commissions, giving a total income from all this kind of work of nearly £5,000, equivalent in purchasing power to several million pounds today. But science was still Hooke's enduring interest, as is shown by the design of one building in London that is undoubtedly his own work, even though the plaque that is on it wrongly attributes it to Wren – the Monument.

The third of Hooke's great achievements resulting from the Fire, the Monument is a hollow memorial pillar near the site of the outbreak of the Fire, 15 feet in diameter and 120 feet high, standing on a square plinth and surmounted by a representation of a flaming urn, giving a total height of 202 feet. This makes it 33 feet taller than Nelson's Column in Trafalgar Square. Made of Portland stone, the Monument was built to commemorate the recovery of the city from the Fire, rather than as a memorial to the Fire itself, so the project only got under way once that recovery was assured. The paperwork for Hooke's design, approved and signed by Wren on behalf of the King, survives, so its authorship is undisputed, although Wren worked with Hooke on practical aspects of the construction, ensuring that it could be used as an astronomical telescope and as a vertical laboratory for carrying out experiments involving gravity. This represented a brilliant application of what ended up as a cost of £13,450 11s 9d from the coal fund – a sum that would never have been devoted to pure scientific research. Lisa Jardine* has said that the Monument 'stands out for the fully realised nature of its double function as both architectural monument and oversized scientific instrument.'

The site for the Monument was available because it was decided

---

* On a Grander Scale.

not to rebuild St Margaret's Church in Fish Street, but to merge the parish with the adjacent St Magnus the Martyr. Permission was formally given to demolish the remains of the old church on 3 August 1670, and on 26 January 1671 the Court of Aldermen approved Hooke's design:

> Upon view of the draught now produced by Mr Hooke one of the Surveyors of new buildings of the Pillar to be erected in memory of the Late dismall Fire the same was well Liked and approved. And it is referred to the said Surveyors to estimate and certify unto thus Court the charge of the said Pillar.

Detailed design work and the legal acquisition of the land by the City took many months, so it was not until October 1672 that orders were issued to enclose the land required for the building site and work could begin. Hooke kept a careful eye on progress, in between his many other commitments, but the main structure was not completed until October 1676, with a modest official opening ceremony on 17 November that year. Sorting out the accounts and settling all the payments to contractors took more months, with Hooke closely involved, but what we are interested in here is how the Monument was intended (and used) for scientific work. To this end, Hooke took meticulous care to ensure that the Monument was built on solid foundations, dug into a bed of gravel six feet thick; that the height of each of the 311 steps leading to a viewing platform at the top, like the staircase of a lighthouse, was accurately monitored at exactly six inches;* and that the pillar was truly vertical with an even interior diameter. No wonder it took so long to build and cost so much. And no wonder it still stands!

Hooke's chief hope was that he would be able to use the vertical

---

* A further thirty-four steps, making a total of 345 (not 365 as is often mistakenly reported), lead from the viewing platform to the very top of the Monument. Being inside the Monument with its narrow winding stair feels a bit like being inside the rifled barrel of a huge gun.

tube of the pillar as a telescope to observe stars directly overhead (at the zenith), and measure the apparent shift in the positions of these stars as the Earth moved around the Sun. This parallax effect is exactly the same as the way in which if you hold a finger out at arm's length and shut each of your eyes in turn the finger seems to move against the background. Hooke had previously tried to measure this by observing from his rooms in Gresham College, using a tubeless 'telescope' made by putting one lens in a circular hole in the roof, with another lens supported directly below it, but the difficulty of lining the lenses up and the vibrations that shook the building made this a hopeless task. The Monument was much better suited to the job, with an objective lens mounted under a hinged lid incorporated into the flaming urn at the top, and a second lens, the eyepiece, in an underground chamber (a large laboratory) below the pillar, where Hooke (or some other observer) could lie on his back to observe. The technique was sound, and astronomical parallax can indeed be measured, but the differences are so small that with the technology available to him even the ingenious Hooke could not detect them – the first measurements of stellar parallax were not in fact made until the 1830s.

Hooke also carried out experiments to measure the air pressure at different heights up the pillar, but these only confirmed what was already known from taking barometers up and down hills, or inside tall buildings such as the old St Paul's Cathedral. Gravity experiments involved dropping objects from different heights inside the pillar and timing their fall, weighing things at different heights, and work with pendulums. No dramatic discoveries were made, but these are all examples of the busy Hooke taking every opportunity, and making new opportunities, to test hypotheses directly by experiment – *nullius in verba*.

Although long misattributed to Wren, in recent times the Monument has often been described as the most visible and appropriate monument to Hooke the architect – at least, the most prominent such memorial in London. But there is something even more impressive, the fourth of his great architectural achievements to which we want to draw attention.

In the course of his work as a surveyor, architect, and supervisor of building work after the Fire, Hooke was responsible for ensuring that the new buildings were structurally sound and capable of bearing the loads put upon them. He became intrigued by the puzzle of finding the best possible shape for an arch, requiring the least amount of building materials for the most strength of the structure. Hooke was the first person to realise that this optimum shape is just the inversion of the curve formed by a chain hanging between two points. Gravity pulling down on the chain produces a tension that shapes the chain into what is called a catenary curve. If you imagine the chain being set rigid and then turned upside down to make an arch, the forces are reversed to become compression forces, the mirror image of the original forces, and this shape, it turns out, provides the greatest strength for an arch – an inverted catenary. With his architectural work now contributing to his scientific work, Hooke described the discovery to a meeting of the Royal in December 1670, although clearly he had been aware of it for some time, and had discussed it with his colleague Christopher Wren. Exactly a year later, on 7 December 1671, he presented a more detailed description, recorded in the Royal Society's archive:

Mr. Hooke produced the representation of the figure of the arch of a cupola for the sustaining of such and such determinate weights, and found it to be a cubico-paraboloid conoid; adding that by this figure might be determined [i.e. solved] all the difficulties in architecture about arches and butments.

At that time, Wren was about to become deeply involved in the project for which he is best remembered, the new St Paul's Cathedral. Before the Fire, he had already been involved in schemes to repair the old St Paul's, including drawing up plans for a new dome based on a double structure, with an inner dome made of brick and an outer dome made of timber covered with lead, to combine strength with an elegant shape viewed from outside. After the Fire, the dream that the old cathedral might be saved persisted among many of the King's advisers, and Wren became increasingly exasperated by being asked to work on proposals that he felt were futile, and for which, in any case, no funds were available. On 24 May 1668, he wrote to the Dean of St Paul's (and later Archbishop of Canterbury), William Sancroft, making it clear that he had had enough, and would not be willing to take responsibility for the complete rebuilding of the cathedral (which it was now recognised was necessary) until the authorities had given him a budget ('silver') to work with. Using an architectural analogy, he said:

> it is silver upon which the foundation of any worke must be first layd, least it sinke while it is yet rising. When you have found out the largeness and security of this sort of foundation I shall presently resolve you what fabrick it will bear.

There were still further delays before the site was cleared and the project for a new cathedral could go ahead, but on 12 November 1673 the King at last issued a 'Commission for rebuilding the cathedral church of St Paul, London'. Wren, of course, was to be its architect. The long saga of the construction need not concern us here (it was not completed, in the sense that the cross was set up, until 1711), but it is noteworthy that Hooke's diary records more than a hundred visits to the site between 1672 (when the demolition of what was left of the old cathedral took place) and the end of 1680, indicating his role as an assistant and adviser to Wren on his 'Cousin's' greatest venture. The most important piece of advice concerned the dome of the cathedral.

At the start of the project, Wren was still thinking in terms of a double dome like those of Florence and Byzantium – no such structure had been attempted in England at that time. But Hooke had a better idea. Thinking about the problem of making the dome as light as possible, he hit on the idea of extending his work on arches. Instead of a single chain dangling between two points, he took a fine mesh, like chain mail, fastened around its rim and hanging to make a curved bowl. Inverting this bowl shape would, just as with the single chain, convert the tension forces into compression forces, and ensure stability to the domed structure. But suppose you didn't want a simple bowl shape. Hooke had the answer. If you added little weights to the hanging mesh, and additional links to the mesh as necessary, you could alter its outline as you required, to achieve the aesthetically pleasing result you sought. Supporting the inverted structure would ensure that it retained its shape with minimum weight. Wren was delighted. On 5 June 1675, Hooke wrote in his diary 'At Sir Chr Wren . . . He was making up my principle about arches and altered his module [model] by it.'

But Hooke did not go public with the idea. Instead, in 1676, in what would otherwise have been a blank space at the end of the printed version of one of his Cutlerian lectures, he added a coded message that reads, when unscrambled, 'As it hangs in a continuous flexible form, so it will stand contiguously rigid when inverted.'

Perhaps he should have made more fuss about his contribution. The resulting dome of St Paul's has been described by Edmund Hambly, one-time President of the Institute of Civil Engineers,* as:

> A masterpiece of structural engineering. It is like an egg-shell in comparison with Brunelleschi's fine dome for the Duomo in Florence.

And it is Hooke's dome, albeit Wren took the idea and made it concrete. The Monument is impressive enough, but the best monument to Robert Hooke the architect is the Dome of St Paul's Cathedral.

---

* See Paul Kent & Allan Chapman.

CHAPTER FOUR

MEANWHILE . . .

What else was Hooke up to in the decade following 1666, the happiest years of his life? At last, we can begin to fill in some details of his domestic and personal life, because many of his letters from the mid- to late 1660s survive, and he started keeping a diary in 1672. This first diary runs until 1683, and a second one starts at the end of 1688 and finishes in August 1693. The material has been dissected and summarised by Lisa Jardine. But Hooke's diary was not a narrative description of his life, like the famous diary of Samuel Pepys; rather, it was a set of brief notes, often written a few days after the events, to provide a reminder to Hooke about what had been going on when. It sometimes lapses into pejorative comments about people, such as his nemesis Henry Oldenburg, but is mostly a factual and straightforward account, even down to details of his sexual activity.

We do not intend to go into great biographical detail, since our main interest is Hooke's scientific work and its significance, but a flavour of Hooke the man may help to bring him out of the shadows. Although required by the rules of the Gresham professorship to remain 'celibate', and living in the college, Hooke had a variety of people sharing his accommodation during his most active years. First, of course, there were the servants. At the time the diary begins, in 1672, the servant was a girl called Nell Young; with scientific detachment, Hooke recorded their sexual encounters in his diary with the astrological symbol for Pisces:)-(.* Presumably there had been other servants on the same intimate terms before the time the diary began. Nell had connections on the Isle of Wight, and may have been the sister of John Hooke's servant. When Nell married and left his employment in the summer of 1673, they remained on good (but seemingly platonic) terms, and she passed on news to him of events on the island. She lived near the Fleet Ditch and took in needlework, including making or mending clothes for Hooke. After Nell left, a succession of less satisfactory servants followed. Bridget, Doll and Bette didn't last long (though long enough for the odd sexual encounter), but in September 1674 Mary Robinson arrived and stayed as his servant, with no mention of sexual shenanigans.

Hooke's household also included lodgers. Richard Blackburne, a Cambridge graduate, was in residence from December 1672 to November 1673, before going to Leyden to study medicine. He was replaced by Harry Hunt, who was a kind of apprentice. Hooke trained Hunt to be his assistant, so successfully that in 1676 he became the Royal Society's 'Operator', at a salary of £40 a year. They remained friends for life. This is the most notable example of the way Hooke was eager to pass on his knowledge and skills to the next generation, seemingly without any jealousy or fear of being superseded. One of his other lodger/assistants was Thomas Crawley.

Two of Hooke's young relatives from the Isle of Wight also came to stay with him, with a view to improving their education and prospects. His niece, Grace, arrived some time in 1672, when she was twelve years old – possibly Nell came with her. In July 1675, Tom Giles (also spelled Gyles), who was the grandson of

---

* Incidentally, Hooke was no astrologer, and wrote disparagingly in his diary (25 November 1678) that it was 'vaine'.

Hooke's uncle (also Thomas Gyles, Hooke's mother's brother), was taken in to be educated with a view to becoming a ship's navigator. We don't know his exact age, but although Hooke described him as lazy and threatened to send him home to the island in March 1677, he softened when the boy burst into tears, and allowed him to stay. When Tom developed smallpox and died in September that year, Hooke called three doctors in to tend to him, and was deeply distressed when their efforts proved worthless.

Hooke's relationship with Grace was deeper, and longer lasting, but would ultimately prove even more distressing. Her story extends into the 1680s and serves as a background to the discussion of Hooke's scientific life in the 1670s and beyond. At the time she arrived in London, her father John, Robert Hooke's brother, was a respected member of the community on the Isle of Wight. He had twice been Mayor of Newport. But he was never good with money, and had already borrowed from his increasingly wealthy younger brother. Robert Hooke, of course, was highly respected in City circles through his work following the Fire, and was on good terms with Sir Thomas Bloodworth, who had been Lord Mayor at the time of the Fire. It seems that a betrothal had been arranged between Grace and the son of Sir Thomas, and she was to be educated and polished to become a suitable consort for the scion of such a family. But that, of course, would involve Grace taking with her a dowry, presumably to be provided by Robert. Early in September 1672, Hooke's diary records 'Bloodworth here, when he resolved to continue to have Grace and to send me his dymands next day.'

There is no hint of any problems with the arrangement until July 1673, when Bloodworth initiated steps to cancel the betrothal. By then, Grace, now thirteen, was back on the Isle of Wight. The legal process of untangling the betrothal was complicated, and not completed (by Hooke, acting on behalf of Grace and her father) until 17 September 1675, when she was fifteen. Meanwhile, Grace had been spending some time on the Isle of Wight, and some in London, where Hooke bought her fine clothes and other presents. In April 1676 she seems to have settled in London and begun once again a more concerted effort at education for life there as a 'Lady', including studying French. Hooke often took her out, and although as a pretty and vivacious young girl she had plenty of other admirers, in October 1677, when Grace was sixteen or seventeen, and he was forty-two, the symbol)-(begins to appear next to her name in his diary.

The age difference between the two would have been unremarkable in those days, but even in Restoration England their relationship, while not illicit in terms of the civil law, could certainly have caused problems with the ecclesiastical authorities. It seems to have been either ignored or accepted by contemporaries (such as Pepys), but in August 1677, perhaps because Hooke felt guilty and intended to end the relationship, Grace went back to the Isle of Wight and Hooke wrote in his diary (10 August) 'Cousen Grace into the Countrey'. Jardine suggests that his use of the word 'cousen' was a reminder to himself that 'their relationship was verging on the improper'. But Grace's return to the island proved a disaster.

Grace was a beautiful seventeen-year-old girl, fashionably dressed, brought up in the ways of Restoration London, and already (thanks to Hooke) sexually experienced. The Governor of the Isle of Wight, Sir Robert Holmes, was a rogue with a roving eye. On 31 October, Hooke heard from Nell the island gossip that 'Sir R. Holmes courting Grace'. 'Courting', of course, was a euphemism. On 3 November, Hooke wrote to his brother 'about Grace and Sir R. Holmes', but the letter has not survived. The next mention of her in the diary, on 26 February 1678, tells us that Grace is confined to bed with 'measles'. And the next day, Hooke's brother John committed suicide.

Any reconstruction of the events leading up to John's death must be speculative, but the most probable explanation is that Grace had been made pregnant by Holmes. John Hooke is known to have suffered from bouts of depression, and in February 1678 he was deeply in debt, not least to Robert. Robert Hooke was always generous with his money, lending to people such as the impecunious John Aubrey with no expectation of ever getting his money back, although he always kept careful records of how much had been lent (he was a meticulous record-keeper in all aspects of his life). He would have regarded the 'loans' to his brother as gifts, but John would doubtless have been ashamed of the need to beg off his brother, and also owed money elsewhere, not least to Newport Corporation. His daughter's pregnancy, removing any prospects of her achieving a good marriage, may have been the last straw.

As if this were not bad enough, there was worse for his family. As a suicide, John Hooke's property would be forfeit to the Crown. Robert Hooke immediately went to Court to beg the King to waive his right to the estate – it is a sign, incidentally, of Hooke's status that he was able, with the help of Wren, to gain almost immediate access to the King – and found that Sir Robert Holmes had already made the same request on behalf of Grace and her mother. It is pretty obvious why Holmes should have felt it his duty to help. The appeal was successful, but Hooke still had to sort out his brother's estate, pay off his debts, and look after Grace, who was now a 'fallen woman' with no prospects except as a servant. The obvious solution was for her to return to London, which she did in May 1678, nine months after her first encounter with Sir Robert Holmes, and become Hooke's official (paid) housekeeper, and unofficial mistress.

Meanwhile, back on the island, Holmes acknowledged that he was the father of a baby girl, who became known as Mary Holmes, and was brought up in his household. Holmes never married, and when he died he left his entire estate to a nephew, on condition that he married Mary, which he did. The union produced sixteen children.* The mother of Mary Holmes has never been officially identified; we leave you to draw your own conclusions.

In London, Hooke and Grace settled into what seems to have

---

* There has been speculation that Hooke was actually the father of Mary, on the grounds that the Pisces symbol appears in his diary alongside Grace's name the day before she left for the Isle of Wight. If so, the sixteen children and their descendants continued Hooke's line, but this is no more than speculation.

been a happy and at least affectionate and, certainly on his part, loving relationship, which forms the backdrop to his life up until 1687, when Grace, still only in her late twenties, died after a short illness. Hooke, who was then fifty-two, was devastated by the loss, and, coupled with other events in the scientific world at that time, this at least temporarily plunged him into a state of gloom. According to Richard Waller, who knew him in later life, he was 'observ'd from that time to grow less active, more Melancholly and Cynical', but this may have been an exaggeration (see Chapter Nine).

But that is getting ahead of our story. In the 1670s, Hooke was certainly active, and by no means 'Melancholly and Cynical'. Hooke's activities were so varied that it is impossible to reconstruct a 'typical' working day (all his days were working days, although New Year's Day was reserved for making up his accounts and taking stock of his life, and Sunday was similarly devoted to administration). But with the aid of 'Espinasse's study of his diaries, and some similar investigation by Steven Shapin,* we can give some idea of just how active he was. Hooke was an early riser, if he had properly been to bed at all, and after a modest communal breakfast with his lodgers or resident assistants before Grace was up and about, he was ready to start work. During the rebuilding of London, this might involve one or more views in the morning, but when this work lessened his mornings were devoted to devising experiments requested by the Royal, preparing lectures, working on his own mechanical contrivances, and maybe architectural work.

In the afternoons, Hooke was out and about as much as his other commitments permitted. He often visited Robert Boyle at Lady Ranelagh's house, where Boyle now had a laboratory. He visited booksellers, bought an inordinate amount of 'medicine' from apothecaries, and frequented makers of good-quality clothes and shoes. Hooke was no dandy, and seems to have dressed plainly (usually in black), but with excellent taste and sparing no expense. At a time when Hooke's protégé Harry Hunt made a decent living as the Royal's operator at £40 a year, Hooke was happy to

---

* See Hunter & Schaffer.

pay one pound a yard for black cloth for a suit, and then to add black ribbon to trim it at sixpence a yard. Black silk stockings cost 11 shillings and sixpence, while a beaver hat cost £3. But his biggest sartorial outlay was on shoes. As he walked everywhere and kept no carriage, he had his shoes made to measure and of high quality. And as well as his book habit, he spent large amounts on scientific equipment. He could well afford it. He was, of course, paid in cash, and from time to time after making up his accounts he would lock some of the money away in a large chest – the usual procedure of affluent people in those days before the advent of secure banks and safety deposits.

On his walks around London, Hooke frequented many of the coffee houses and taverns, where he met – sometimes by appointment, sometimes by chance – friends and colleagues from the worlds of science, surveying or the mechanical arts. These meetings were a mixture of work and pleasure. As we have noted, surveying work was sometimes discussed over a coffee; sometimes the subject under discussion might be flying machines, or gravity. The diary shows that between 1672 and 1680 Hooke visited some sixty coffee houses and more than seventy taverns, but until 1677 his favourite was Garraway's, in Exchange Alley, close to the archetypal Pasqua Rosa's. This was just five minutes' walk from his rooms. But in July 1677 a new coffee house, Jonathan's, opened just around the corner from Garraway's. Hooke and his friends soon switched their allegiance there. For more than a decade, Jonathan's was a centre of scientific discourse; it later became the haunt of financial speculators and stock dealers. But during Hooke's time, when the Royal Society itself became irksome for one reason or another, Hooke and his like-minded scientific friends would get together as an informal science club to meet there in more congenial surroundings.

Hooke's social life largely took place outside his rooms at Gresham College, at places such as Jonathan's or on visits to people such as Boyle and Wren. Indeed, Hooke's diary contains so many references to meetings with Wren that it is one of the prime sources used by Wren scholars. Hooke was a popular man about town, but seldom entertained at home; the convention of the time was that those regarded as lower on the social scale should call on those higher up, but not the other way around. And in many ways Hooke was a stickler for convention. Even in his private diary he referred to people by their proper titles – Lord Brouncker, Mr Boyle, and so on. Wren was Dr. Wren until receiving his knighthood, after which he is Sir Ch. Wren.* Hooke knew his place, and was happy to defer socially to his superiors, but equally insisted on the respect due to his own status. Although he did not entertain in the way his social superiors did, he did have a few visitors – including John Aubrey, who sometimes needed a place to stay, and on a more regular basis Theodore Haak, a German-born scientist and FRS with whom Hooke regularly played chess until Hack died in 1690. Clearly, Hooke was (at least until Grace died) gregarious, sociable and well liked, as well as being highly respected. Jardine has described him as:

> sociable, obliging, enthusiastic, and, in terms of the organisation of his efforts, an inveterate optimist. Generous to a fault, he tried valiantly to give adequate attention and time to each of the projects he undertook.

The only two people Hooke ever fell out with, in each case with good reason, were Newton and Oldenburg, although naturally he had scientific disagreements, as opposed to personal feuds, with others; these included, as we shall see, the astronomer Johannes Hevelius.

Some days Hooke had to give his Gresham or Cutlerian lectures, and, although these were often poorly attended, the need to prepare them had the happy consequence for us that written records survive of his thoughts on diverse subjects. When the Royal met (at first on Wednesdays, later Thursdays), he had to carry out demonstrations, and after the meetings he would usually

---

* The exception is Henry Oldenburg, who is usually simply referred to as 'Oldenburg', but occasionally as 'lying dogg Oldenburg', 'villain Oldenburg' or 'huff Oldenburg'.

be among a group of Fellows who adjourned to a coffee house to burn the midnight oil with philosophical discussions. Even then he would not be finished. Especially after 1674, when he had a turret built on to his rooms at the College, Hooke would stay up until the small hours, and sometimes all night, making astronomical observations. Waller tells us that Hooke was frequently 'continuing his Studies all Night, and taking a short Nap in the Day'.

How was it possible for one man to do all this? Partly because Hooke was, as we have said, a chronic insomniac and workaholic. But also because he dosed himself with a fearsome quantity, and even more fearsome quality, of medication for real or imagined illnesses, and drank copious amounts of coffee.

Lucinda McCray Beier has described Hooke as 'the great medicine-taker of seventeenth-century England'.* His diary seldom describes any specific ailment, simply remarking that he is ill. But he does mention giddiness, headaches, dizziness and palpitations (all of which, as Jardine has commented, 'may have been side-effects of the preparations he was consuming'†) and seems to have suffered from noises in the head, presumably tinnitus, which he obscured by singing while he worked. He seems to have regarded his bodily ills as something of an opportunity for scientific studies of the value of various 'cures', but also took medicines not to treat existing symptoms but in the hope of preventing symptoms that he imagined might otherwise be felt. Even though he actually suffered no major illness during the period when he kept the diary, he would take emetics and purgatives to cleanse his body, sometimes with the result (for example, on 4 September 1672) that he was 'disordered somewhat by physic'. He would drink Dulwich water, a popular 'health' drink from a spring at that location, by the pint, and on occasion tried 'syrup of poppies' to help him sleep, resulting in sweaty nights filled with 'wild, frightful dreams'. For a time, he tried a

---

* Hunter & Schaffer.
† Bennett, Cooper, Hunter & Jardine.

preparation containing iron and mercury to see what effect it had. He tried tobacco to see if it had medicinal properties, and noted the effects of coffee, chocolate, tea and wine on his body. A not untypical entry reports on 16 February 1673, after using 'Andrews cordial', that this:

> brought much slime out of the guts and made me cheerful. Eat dinner with good stomach and pannado at night but drinking posset upon it put me into a feverish sweat which made me sleep very unquiet and much disturbed in my head and stomach. Taking sneezing tobacco about 3 in the morn clear my head much and made me cheerful afterwards. I slept about 2 hours, but my head was disturbed when I waked.

And on 1 August 1675:

> Took volatile Spirit of Wormwood which made me very sick and disturbed me all the night and purged me in the morning. Drank small beer and spirit of Sal-amoniack. I purged 5 or 6 times very easily upon Sunday morning. This is certainly a great Discovery in Physick. I hope that this will dissolve that viscous slime that hath soe much tormented me in my stomach and guts.

Hooke, indeed, took so much medication that it was news when he did not. On 3 August 1673, the diary records 'took no physick'. Beier emphases that there was nothing unusual about the kinds of 'physick' that Hooke took, only a few of which we have mentioned here; his contemporaries swallowed much the same preparations. But 'few people took as much physic as often as did Hooke'. Although he was happy to take medical advice from friends, strangers and supposed wise women, he also consulted the best physicians in London more often than most of his contemporaries. This may have been partly because of the circles he moved in, which gave him access to the medical elite, but it may also reflect his scientific interest in medicine and its effect on the human body, especially his body. Having tried everything, he was certainly well placed to pass on medicinal advice to others, which he happily did.

One of the effects of all this physic taking, Jardine suggests, was that it 'excites the mental faculties, producing a clarity which is conducive to slightly fevered intellectual activity of the kind Hooke needed in order to cope with his chronic burden of over-work and the competing demands of clients and employers.' He operated, she says, 'under the permanent influence of stimulants'. If so, it may explain his physical decline in later years. But it could also explain how he achieved so much in so many areas, the theme to which we shall now return.

One of the first significant post-Fire experiments that Hooke was involved in resulted from criticism of his suggestion that breathing supplies the body with the same substance as 'that which is fixt in Salt-peter', and is essential for life. In a demonstration to the Royal on 10 October 1667, Hooke and a colleague, the leading anatomist Dr Richard Lower, made small holes in the lung wall of a dog so that its lungs could be kept permanently inflated, with air pumped in from a bellows through one hole and out through the other. The dog 'lay still as before, his eyes being all the time very quick, & his heart beating very regularly'. This established that the supposed pumping action of the lungs was not important in itself, but was simply the way the body took in fresh air and expelled stale air. In further experiments a little later, stale air from the lungs was recycled until the dog seemed near to death, but the animal recovered when allowed to breathe normally.

Lower developed these ideas in further studies, which formed the core of his influential book *De Corde* (*On the Heart*), published in 1669, where he gave Hooke due credit. But this was the last time Hooke was involved in vivisection, having made a key contribution to the development of an understanding of the way the heart and lungs contribute to the workings of the body.

Apart from his relative squeamishness discouraging him from further work of this kind, Hooke was busy on other things. His

work on pendulums and timekeeping devices never stopped, and
we have a detailed description of one contribution thanks to a
visitor from Florence, Lorenzo Magalotti, who was at the Royal
in February 1668 and reported:

> We also saw a pocket watch with a new pendulum invention.
> You might call it with a bridle, the time being regulated by
> a little spring of tempered wire, which at one end is attached
> to the balance-wheel, and at the other to the body of the
> watch. This works in such a way that if the movements of
> the balance-wheel are unequal, and if some irregularity of
> the toothed movement tends to increase the inequality, the
> wire keeps it in check, obliging it always to make the same
> journey. They say that if you keep it hung up, the invention
> works well and that it corrects the errors of the movement
> as well as a pendulum, but that if you carry it in your pocket
> the temper of the wire changes in accordance with the
> temperature of the body, and getting softer, allows the
> balance-wheel to turn with more freedom.*

There are two noteworthy things about this description of
Hooke's invention. First, because it is affected by temperature, it
does not solve the longitude problem. And, curiously, no record
of this demonstration survives in the official Royal Society records,
which at the time were the responsibility of Henry Oldenburg,
whom Hooke suspected, on the basis of very persuasive evidence,
of being in cahoots with Christiaan Huygens in an attempt to get
patent rights for the Dutchman's chronometer. Indeed, as
'Espinasse has pointed out, this was not the first time that
Oldenburg 'omitted to record the demonstrations' of Hooke's
timekeepers. Coincidence?

One seemingly minor aspect of Hooke's activities in the late
1660s is sometimes overlooked, but shows great insight. He tried

---

* See R.D. Waller, in 'Lorenzo Magelotti in England, 1668–9', *Italian Studies*, i. 2, 1937–8.
Also quoted by 'Espinasse.

mixing gold and lead to see if they would combine to form a substance denser than either of them. This was not the work of an old-school alchemist, but was based on the idea that there might be gaps between the particles of each element into which particles of the other substance might penetrate, filling in the gaps. This wasn't quite the concept of atoms, but it came mighty close.

Another development in 1668 had more immediate implications, and would eventually involve the other main character in our story, Edmond Halley. In June that year, Oldenburg received a request from Johannes Hevelius, a German based in Danzig and Europe's leading astronomer of the time. Hevelius, who had been born in 1611, was an astronomer of the old school. He had a superb observatory equipped with the finest quadrants and other 'open sight' instruments, and large, but not quite so fine, telescopes. He made his measurements of stellar positions, and (for example) the way a comet moved past the background stars using 'open sights', and was convinced that this was more accurate then the use of the telescopic sights attached to such instruments, which Hooke and Wren had pioneered and which Hooke had described in a letter to Hevelius. Nevertheless, he wanted to improve his telescopic observations for other tasks, such as mapping the Moon, and asked Oldenburg to obtain the best telescope he could for him, while also making the contradictory demand that it should not be expensive. Oldenburg passed on the task to Hooke, who found a suitable instrument and sent it off to Danzig. This initiated further correspondence between Hevelius and Hooke, with Hevelius insisting that his open sights method was more accurate than Hooke's telescopic sights observations. Neither man would give way, and there the matter rested for the time being.

Hooke also corresponded with a young astronomer in Derby, John Flamsteed, who was eager to know how Hooke was able to grind the lenses for his telescopes, and became increasingly frustrated by Hooke's reluctance to pass on what he regarded as trade secrets. Flamsteed too features later in our story.

In November 1668, Hooke demonstrated to the Royal an experiment that he had first carried out in October 1666, but which had received little attention in the aftermath of the Fire. It involved three balls of equal weight suspended as pendulums, initially touching one another. When one ball was pulled to one side and allowed to swing down on the other two, it stopped, the middle ball was seemingly unaffected, and the ball on the other side swung up as if completing the pendulum swing of the first ball. Sounds familiar? It was the prototype of the executive toy now known as Newton's Cradle. It has nothing to do with Newton, but demonstrates Hooke's discovery 'that no motion dies, nor is any motion produced anew', as the Society records put it. Developing these ideas, Hooke was soon able to show that in order for an impact to double the speed of an object moved by an impact, the force on it had to be quadrupled. While Hooke's terminology is hard to unravel today, he clearly had some idea of the property we now call momentum. And early in 1671 he was studying the way different sounds made patterns in a shallow dish of flour, which he hoped 'might much contribute to the explication of the nature of the internal motion in bodies'. Meanwhile, during these years, as we have seen, as well as his work as surveyor and architect, he was developing his ideas about, and lecturing on, earthquakes and their implications.

In the midst of all this, some claims made by an upstart young Fellow of Trinity College, Cambridge, caused Hooke some irritation, fanned into a more serious dispute by Oldenburg's malicious interference.

Isaac Newton's study of light, stimulated by his reading of Hooke's *Micrographia*, had led him to the realisation that the passage of white light through a prism or other glass separates the light into its constituent colours. This explained why the images obtained using refracting telescopes were edged with coloured fringes. Although modern telescopes can get round this problem by using compound lenses made of different materials, so that the effect of one part of the lens cancels out the effect of the other, this requires a technology far more advanced than that of the 1660s. Newton realised that the problem would not arise in a reflecting telescope using a curved mirror, rather than a lens, to concentrate the light. The obvious problem with such a device is that in order to see the image in the mirror the observer would have to look down the tube, blocking out the light coming into the instrument.

Newton found a simple solution to this problem. He put a small flat mirror near the mouth of the telescope tube, tilted at an angle of 45 degrees, so that the focused reflected light from the main mirror bounced off the flat mirror and out through a hole in the side of the tube. The observer could look in through the side of the telescope to see what was overhead. Although Newton did not know it, the idea was not new. Back in the 1550s the English surveyor Leonard Digges came up with essentially the same design as a spin-off from his work with theodolites, but kept it secret so that he would have an advantage over rival surveyors. As far as we can tell, it was based on what is now known as the Newtonian design. Digges probably also invented a refracting telescope, but that is not certain. Nor was Newton the first person thinking about reflecting telescopes in the mid-seventeenth century. The Scot James Gregory had come up with a different design in 1663. In the Gregorian design, the main concave mirror reflects and focuses light on to a smaller concave mirror in the middle of the tube, from where it goes back down through a hole in the centre of the main mirror to the observer.

Gregory was more of a mathematician than a practical man, so the first telescope using the design was made for him by Richard Reeve, London's leading instrument maker, in 1664. In spite of Reeve's skill, the telescope did not prove very satisfactory. Hooke tested it on behalf of the Royal Society, and makes passing mention of reflecting telescopes in *Micrographia*, but found that he could obtain a sharper image with his reflecting telescopes, even with the irritation of the coloured images, so he gave up any attempt to develop the idea further. Gregory himself left to further his career in Italy.

Newton's first telescope was made in 1667 or 1668, shortly after he read *Micrographia*, but he did not publicise it. This was typical: Newton was a rather odd person, somewhat reclusive, who tended not to bother telling people about his work unless he was prompted to do so.* But by 1671 visitors to Cambridge who had seen the telescope (or possibly a second one made to the same design) had spread news of it as far as London, where the Royal asked to see it. At the end of that year, Newton's Cambridge colleague Isaac Barrow took the instrument to Gresham College and showed it to the Fellows. They were sufficiently impressed to appoint a committee consisting of Hooke (of course), Wren, Brouncker, Moray and Neile to test it. They found that, although Newton's telescope was only six inches long, it produced a magnification of thirty-eight times, better than a much larger refractor. On the strength of this, Newton was elected as a Fellow on 11 January 1672, and gave the reflector to the Royal. Newton was twenty-nine, and this was essentially the first time anyone had heard of him; Hooke was thirty-six and established as the greatest scientist in Britain, the 'go-to man' of the Royal Society.

Newton's arrival on the scene had two effects on Hooke's scientific life. The first was to revive his interest in reflecting telescopes, no doubt partly in irritation at the adulation Newton had received. In fact, the shine soon went off Newton's invention – literally. The metal mirror, or speculum, tarnished, and was difficult to keep in shape. By the end of January 1672, Hooke had already started work on what he intended to be a bigger and better reflector, and this continued throughout the year, assisted by one of the London craftsmen, Christopher Cock, but with only limited success. The problems of tarnishing and distortion persisted, even with steel mirrors. And at the same time, Hooke was involved in architectural work, including the College of Physicians, and City work, especially the Fleet Ditch. It wasn't until March 1673 that the idea Hooke needed was put forward by Gregory in a letter to the Royal. Gregory had the brilliant

---

* Newton's oddities may have been related to his upbringing. His father died before Newton was born, his mother remarried when he was three, and he was sent to live with her elderly parents. He developed as a solitary man who had few close friends. He was also a religious nonconformist, who risked losing his Cambridge position if this became known, and probably a closet homosexual, another inducement to secretiveness.

idea of making the concave speculum out of glass with a mirrored metal rear surface, so that the shiny metal surface itself would never come into contact with air and would not tarnish. Just under a year later, on 5 February 1674, Hooke presented a telescope based on this idea to the Royal. It was not only the first practical Gregorian telescope, but it was the first practical astronomical reflector, given the problems with Newton's metal mirror design. And Hooke's application of Gregory's idea of a mirrored glass speculum has continued to be used in great astronomical telescopes up to and beyond the Hubble Space Telescope. But by 1674, the other effect of Newton's emergence from the shadows was beginning to colour Hooke's scientific life.

In order to put the feud with Newton in perspective, we need to look first at the mischievous activities of Henry Oldenburg. Like Hooke, Oldenburg had been a member of the Boyle household, having been a tutor to Lady Ranelagh's son, Richard Jones. He accompanied Richard to Oxford in 1656, and even registered as a student but with no intention of taking a degree. He joined Boyle's circle of scientific friends in Oxford, but more as an observer and enthusiast than as an active scientist. Although he was not himself a significant scientist, as the first Secretary of the Royal he played a major part in its initial success, and was diligent – sometimes too diligent – about communicating with scientists across Europe. He also founded (in 1665) and published (for its first twelve years) the *Philosophical Transactions* of the Royal Society, the first scientific journal in the world, which is still going today.

There is no specific reason known why Oldenburg should have taken a dislike to Hooke, but a natural supposition is that he became jealous of Hooke's success. Oldenburg was a would-be scientist, who had known Hooke as Boyle's mere assistant and saw him rise to heights that Oldenburg himself could never achieve. Whatever the reasons, the antagonism was beginning to show by March 1666, when Boyle objected to Oldenburg's failure to give Hooke due credit for his depth-sounding device in an article Oldenburg wrote for the *Philosophical Transactions*. Since

Boyle knew both men so well, the criticism has to be taken seriously. It is also now well documented that, as Hooke suspected, Oldenburg passed on details of Hooke's work on chronometers to Huygens. Jumping ahead a little, in 1676 Oldenburg published in the *Philosophical Transactions* a translation of a letter from Huygens describing his latest pocket watch:

> the invention consists of a spring coiled into a spiral, attached at the end of its middle to the arbor of a poised, circular balance which turns on its pivots; and at its other end to a piece that is fast to the watch-plate. Which spring, when the Ballance-wheel is once set a going, alternately shuts and opens its spires, and with the small help it hath from the watch-wheels, keeps up the motion of the Ballance-wheel, so as that, though it turn more or less, the times of its reciprocations are always equal to one another.

Interesting to compare that with the description of Hooke's watch from 1668 (see page 23), mention of which Oldenburg 'forgot' to include in the records of the Royal. It may or may not be that Huygens hit on similar ideas to Hooke independently, but it is certain that Hooke had the ideas first, and was done out of due credit by the machinations of the Secretary of the Royal. Indeed, in 1675 Hooke had a watch made to his own design by Thomas Tompion, a leading London watchmaker, and presented it to the King. It was inscribed in Latin, 'Hooke invenit 1658. Tompion fecit 1675.'* In his diary, Hooke wrote: 'With the King and shewd him my new spring watch, Sir J. More and Tompion there. The King most graciously pleasd with it and commended it far beyond Zulichems.'† The same year, he wrote to Aubrey: 'I have many things which I watch for an opportunity of Publishing, but not by the Royal Society. Oldenburg his snares I will avoid if I can.' But in the end, neither Hooke nor Huygens was granted a patent.

---

* 'Invented by Hooke 1658. Made by Tompion 1675.' Even taking the 1658 date with a pinch of salt, it was Hooke's invention.

† Huygens was formally referred to as 'Christiaan Huygens of Zulichem'.

In fact, Oldenburg 'had previous', as they say in TV detective stories, although he singled Hooke out for special attention. Putting it charitably, Jardine says that he had the 'instinct of the adept publicist', eager to whip up controversy and thereby 'spice up exchanges of letters which would later find their way into his published *Philosophical Transactions*.' She cites the example of a letter from the French scientist Adrien Auzout, sent to Hooke via Oldenburg. Oldenburg highlights points for Hooke to respond to, with comments such as 'What say you to this?' and 'A handsome sting again will be necessary.' And in correspondence from Oldenburg to foreign scientists such as Huygens he describes Hooke as 'a man of unusual humour' and suggests that his genius (which even Oldenburg cannot deny) comes close to crossing the line into madness. On 20 May 1677, Hooke had reached breaking point, and considered resigning from the Royal; he wrote in his diary 'Saw the Lying Dog Oldenburg's transactions. Resolved to quit all employments and to seek my health.' But in the end he decided to stick it out, and not long afterwards Oldenburg died. It is against that background that we can return to our story of what happened when Newton, encouraged to come out of his shell by the enthusiastic response of the Royal to his telescope, was persuaded to send them a letter detailing his ideas concerning light and colour.

This was in 1672, when Hooke was heavily occupied with his work as surveyor and architect, and the Royal had not yet been able to return to Gresham College. Newton, by contrast, resided in the proverbial academic ivory tower, with nothing to distract him from natural philosophy, unless he chose to do something else.*

Hooke was the leading expert on light and optics, and made a careful study of Newton's letter before reporting back to the Royal. The key difference between Hooke's model of light and Newton's model was that Newton regarded white light as being

---

* Among the things Newton did choose to waste his time on were alchemy and slightly barmy religious studies, but we need not go into detail here except to note that if anyone crossed the line between genius and madness he did, at least some of the time.

a mixture of colours even when it was not passing through the glass of a prism (or lens), while Hooke thought that the colours did not exist until the light interacted with the glass. Newton also favoured the idea that light was a stream of particles, but Hooke thought of it as a wave. Even so, Hooke was willing to accept that Newton's hypothesis worked as a possible description of what happened to light when it passed through a prism, but he suggested that it was no better than other hypotheses, in particular his own. He granted that Newton's hypothesis was 'very subtill and ingenious' but concluded 'I cannot think it to be the only hypothesis.' Here is a flavour of Hooke's letter:

But why there is a neccessity, that all those motions, or whatever else it be that makes colours, should be originally in the simple rays of light, I do not yet understand the necessity of, no more than that all those sounds must be in the air of the bellows, which are afterwards heard to issue from the organ-pipes; or in the string, which are afterwards, by different stoppings and strikings produced; which string (by the way) is a pretty representation of the shape of a refracted ray to the eye; and the manner of it may be some-what imagined by the similitude thereof: for the ray is like the string, strained between the luminous object and the eye, and the stop or fingers is like the refracting surface, on the one side of which the string hath no motion, on the other a vibrating one. Now we may say indeed and imagine, that the rest or streightness of the string is caused by the cessa-tion of motions, or coalition of all vibrations; and that all the vibrations are dormant in it: but yet it seems more natural to me to imagine it the other way.

These were far from being unreasonable comments, even if Hooke's tone tended to be rather condescending. Oldenburg sent a copy of Hooke's critique to Newton, but declined to publish it alongside Newton's own letter in the *Philosophical Transactions*. Hooke repeated some of Newton's experiments for the Royal, in particular the one that showed that a rainbow pattern of colours produced when white light passed through a prism could be recombined to make white light by a second prism. But he stuck firmly to his argument that waves of white light could be made up from many different waves, just as the sound produced by a musical instrument could be a combination of different sounds. In this, we now know, he was more right than Newton.

Prompted by Oldenburg, Newton, who was in any case a prickly personality who could not bear criticism, hit back in a long letter addressing all of Hooke's points in turn, and claiming that he found it easier to imagine white light being split into colours and recombined if it was made of different coloured particles. And he pointed out that light, unlike waves, travels in straight lines.

Newton seems to have got more and more angry as he drafted and redrafted the letter. Many copies survive, and as he 'improved' the document he added more and more references to Hooke, making them increasingly offensive. As Newton's biographer Richard Westfall has put it, Newton 'virtually composed a refrain on the name Hooke'.

Oldenburg was delighted. He read out the juicy bits of the letter to the Royal on 12 June 1672, and published the whole diatribe in his journal, without giving Hooke the right of reply. Feeling that it was futile to take up the issue with Oldenburg, Hooke wrote to Lord Brouncker* to express his distress at having provoked so violent a reaction. 'I was soe far from imagining that Mr. Newton should be angry that I cannot yet believe that he is', he wrote; in matters of philosophical discussion 'a freedome & liberty of Discoursing and arguing ought to be Tollerated', and he claimed that he would be happy to see his own hypothesis disproved if that were to be the case. All he sought was the truth about nature.

But the most remarkable thing about this letter, which is not as widely known as it deserves to be, is that he goes on to describe a crucial experiment that he had demonstrated to Brouncker.

---

* It is not clear whether he ever sent the letter, but a draft was found in his papers.

Referring to a narrow beam of sunlight obtained by making a small hole in a window blind, he says:

> By placing the Edge of a razor in the cone of the Suns radi-
> ation at a pretty distance from ye hole . . . and by holding
> a paper at some Distance from the razor in the shadow
> thereof, your Lordship plainly saw, that the Light of the sun
> Did Deflect very deep into the shadow.

This was proof, in 1672, that light travels as a wave. Like waves on the sea bending round an obstruction such as a rock, light waves bent around the edge of the razor and into the shadow. But it was not until the early nineteenth century that the experiments of Thomas Young and Augustin Fresnel overthrew the 'corpuscular theory' of light, which had held sway for nearly 150 years almost solely on the basis of Newton's reputation. By the end of the seventeenth century, his status as a giant of science (*the* giant of science) was so great that anything he had said 'must be true' in the eyes of lesser mortals.

Hooke was not the only one to take issue with Newton about his light hypothesis. Huygens read Newton's letter in the *Philosophical Transactions*, and wrote to him with some comments and criticisms of his own (Huygens was a leading proponent of the wave model of light). He received such an unpleasant reply that he told Newton 'seeing that he maintains his doctrine with some warmth, I do not care to dispute', and gave up the correspondence. For his part, Newton, unable to cope with the kind of frank and fearless discussion that characterised scientific debates of the time, withdrew back into his ivory tower, threatened to resign from the Royal, and laid low for the next couple of years. Hooke carried on as usual, working away at a variety of projects including a calculating machine (based on a design by the German mathematician Gottfried Wilhelm Leibniz*) that used geared wheels to carry out multiplication to twenty places, but which proved too complicated to be practical.

---

* Someone else with whom Newton would later have a furious argument.

Newton had calmed down enough by 1675 to attend several meetings of the Royal, where among other things he saw and heard Hooke describe the diffraction experiment mentioned in the letter to Brouncker. This was probably the first time he had aired the idea in public. Newton was not impressed, and dismissed it as only a new kind of refraction, to which Hooke replied that even if that were the case, it was indeed new.

Early in December 1675, Newton sent Oldenburg a long letter (a scientific paper in modern terminology) on light, which the Secretary read out to the Royal in two parts on successive weeks. As well as setting out his stall and once again responding to Hooke's arguments, Newton included a sneering reference to Hooke's razor experiment, claiming that it had previously been described by the Italian Francesco Grimaldi, and others. Even Newton was decent enough to add 'I make no question but Mr Hook was the Author too', meaning that Hooke made the discovery independently, but the arch-stirrer Oldenburg left that sentence out of his presentation.

Now Oldenburg fanned the flames of the dispute, or, as Hooke would later put it, he 'kindled cole'.

Oldenburg told Newton (untruthfully) that Hooke had claimed that essentially everything described in Newton's letter was contained in *Micrographia*, which Newton 'had only carried farther in some particulars'. Until that point, Newton seems to have been willing to acknowledge his genuine debt to Hooke. In the second part of his long paper, he had said that *Micrographia* contained 'very excellent things concerning the colours of thin plates, and other natural bodies, which I have not scrupled to make use of so far as they were for my purpose.' But in response to Oldenburg's mischief-making, at the end of 1675 and early in 1676 he sent two further letters to the Secretary, saying that Hooke's discussion of light and colour was not original, but largely lifted from Descartes, and that nothing Newton had borrowed from *Micrographia* was actually Hooke's own work. Oldenburg read out the attack on Hooke to the Royal on 20 January 1676, without anyone, least of all Hooke, being warned in advance. Hooke realised that Newton was being manipulated by Oldenburg, and wrote in his diary that 'Oldenburg kindle cole.' He immediately wrote to Newton to soothe him, initiating an exchange of letters so important (and so often misunderstood) that it is worth repeating in full.*

These to my much esteemed friend, Mr. Isaack Newton, at his chambers in Trinity College in Cambridge.

Sir., – The hearing a letter of yours read last week in the meeting of the Royal Society, made me suspect that you might have been some way or other misinformed concerning me; and this suspicion was the more prevalent with me, when I called to mind the experience I have formerly had of the like sinister practices. I have therefore taken the freedom, which I hope I may be allowed in philosophical matters to acquaint you of myself. First, that I doe noe ways approve of contention, or feuding or proving in print, and shall be very unwillingly drawn to such kind of warre. Next, that I have a mind very desirous of, and very ready to embrace any truth that shall be discovered, though it may much thwart or contradict any opinions or notions I have formerly embraced as such. Thirdly, that I do justly value your excellent disquisitions, and am extremely well pleased to see those notions promoted and improved which I long since began, but had not time to compleat. That I judge you have gone farther in that affair much than I did, and that as I judge you cannot meet with any subject more worthy your contemplation, so I believe the subject cannot meet with a fitter and more able person to inquire into it than yourself, who are every way accomplished to compleat, rectify, and reform what were the sentiments of my younger studies, which I designed to have done somewhat at myself, if my other more troublesome

---

* Source: David Brewster, *Memoirs of the Life, Writings, and Discoveries of Sir Isaac Newton*, vol. 1 (Edinburgh: 1855).

employments would have permitted, though I am sufficiently sensible it would have been with abilities much inferior to yours. Your design and mine are, I suppose, both at the same thing, which is the discovery of truth, and I suppose we can both endure to hear objections, so as they come not in a manner of open hostility, and have minds equally inclined to yield to the plainest deductions of reason from experiment. If, therefore, you will please to correspond about such matters by private letters, I shall very gladly embrace it; and when I shall have the happiness to peruse your excellent discourse, (which I can as yet understand nothing more of by hearing it cursorily read,) I shall, if it be not ungrateful to you, send you freely my objections, if I have any, or my concurrences, if I am convinced, which is the more likely. This way of contending, I believe, to be the more philosophical of the two, for though I confess the collision of two hard-to-yield contenders may produce light, [yet] if they be put together by the ears by other's hands and incentives, it will [produce] rather ill concomitant heat, which serves for no other use but . . . . . . kindle – cole. Sr, I hope you will pardon this plainness of, your very affectionate humble servt,

Robert Hooke

This could hardly be more conciliatory, and in the genuine circumstances of Hooke's busy life we can accept his wish to make it clear that in other circumstances he might have taken things further himself. At first sight, Newton's reply is equally conciliatory.

'Cambridge, February 5, 1675–6.

DR. Sir, – At the reading of your letter I was exceedingly pleased and satisfied with your generous freedom, and think you have done what becomes a true philosophical spirit. There is nothing which I desire to avoyde in matters of philosophy more than contention, nor any kind of contention more than one in print; and, therefore, I most gladly embrace your proposal of a private correspondence. What's done before many witnesses is seldom without some further concerns than that for truth, but what passes between friends in private, usually deserves the name of consultation rather than contention; and so I hope it will prove between you and me. Your animadversions will therefore be welcome to me; for though I was formerly tyred of this subject by the frequent interruptions it caused to me, and have not yet, nor I believe ever shall recover so much love for it as to delight in spending time about it; yet to have at once in short the strongest objections that may be made, I would really desire, and know no man better able to furnish me with them than yourself. In this you will oblige me, and if there be any thing else in my papers in which you apprehend I have assumed too . . . . . . If you please to reserve your sentiments of it for a private letter, I hope you [will find that I] am not so much in love with philosophical productions, but that I can make them yield. . . . . . But, in the mean time, you defer too much to my ability in searching into this subject. What Descartes did was a good step. You have added much several ways, and especially in considering the colours of thin plates. If I have seen farther, it is by standing on the shoulders of giants. But I make no question you have divers very considerable experiments beside those you have published, and some, it's very probable, the same with some of those in my late papers. Two at least there are, which I know you have often observed, – the dilatation of the coloured rings by the obliquation of the eye, and the apparition of a black spot at the contact of two convex glasses, and at the top of a water-bubble; and it's probable there may be more, besides others which I have not made, so that I have reason to defer as much or more in this respect to you, as you would to me. But not to insist on this, your letter gives me occasion to enquire regarding an observation you was propounding to me to make here of the transit of a star near the zenith. I came out of London some days

sooner than I told you of, it falling out so that I was to meet
a friend then at Newmarket, and so missed of your intended
directions; yet I called at your lodgings a day [or] two before
I came away, but missed of you. If, therefore, you
continue . . . . . . to have it observed, you may, by sending
your directions, command . . . . . . your humble servant,

'Is. Newton'

There are, though, grounds for suspicion about just how
friendly Newton was being. The sentence 'If I have seen farther,
it is by standing on the shoulders of giants' is at first sight a
repetition of a classical example of false modesty dating back at
least to first-century Rome. Newton was well aware of Hooke's
small physical stature, and at the very least it was tactless to
include the words; at worst, some historians have argued, it was
a deliberate insult, implying 'I have no need to steal ideas from
a little runt like you'. And it is certainly not genuine modesty!
Intentional or not, it is highly likely that Hooke, who was sensitive about his size, would have taken it as an insult.

After this exchange, Newton went back into his shell, and Hooke
was busy with architectural work. But around this time two aspects
of Hooke's life came into conjunction, when his architectural work
brought him into contact with a young man who would become
a friend for life, and would be instrumental in elevating Newton
to the status of a giant in whose shadow they lived.

In March 1675, Charles II appointed John Flamsteed as his
'astronomical observator', a term soon replaced by the title
Astronomer Royal, which we shall use throughout. This wasn't
just a result of the King's rather dilettante interest in science, but
related to the importance of navigation: the Astronomer Royal's
main task, at first, was to carry out a survey of the northern skies,
mapping the positions of the stars using telescopic sights, to improve
upon the survey made by Tycho Brahe in the sixteenth century
using open sights. At first Flamsteed was an 'observator' without
an observatory, but something was already being done about that.

Hooke was closely involved in all this. He was in correspondence with Hevelius about the relative merits of open sights and telescopic sights, and he had developed the best astronomical surveying instruments in England, possibly in the world. It was the success of Hooke's instruments that persuaded Charles that it would be worth the expense of building an observatory, and in September 1674 Hooke and Sir Jonas More (or Moore), who was Surveyor-General of the Ordnance and a good friend of Hooke, were sent to assess the suitability of a former theological college in Chelsea as a home for the observatory. Moore offered to contribute £250 to the cost if the project went ahead. When Flamsteed heard about the project, he began lobbying for the job of running the observatory, buttering up Moore in a series of letters. At first, these included criticism of Hooke (whom Flamsteed saw as a rival for the post) for his secrecy, but when he realised the relationship between Hooke and Moore was friendly, he changed his tune. Flamsteed sent Hooke the present of a firkin of ale, and wrote to Moore 'for the futur my opinion of him shall be much more charitable.' The lobbying worked, and it was largely on Moore's recommendation, with the acquiescence of Hooke, that Flamsteed became the first Astronomer Royal.

Within a week of Flamsteed's appointment, at the end of March 1675 he received a letter from an eighteen-year-old Oxford undergraduate, Edmond Halley, describing some of his own observations, which disagreed with published tables of data, and asking Flamsteed's advice. It turned out that Halley was right and the tables were wrong; by the summer, Halley was acting as Flamsteed's unpaid assistant on visits to London. But those visits did not involve the old theological college in Chelsea.

A committee that included Hooke and Wren decided that the best site for the observatory would be at Greenwich, and that it should be an independent institution, not under the auspices of the Royal Society. On 22 June, the Royal Warrant, which established the Royal Greenwich Observatory, was issued. It begins:

Whereas, in order to the finding out of the longitude of

places for perfecting navigation and astronomy, we have resolved to build a small observatory within our park at Greenwich, upon the highest ground, at or near the place where the castle stood, with lodging rooms for our astronomical observator and assistant, Our Will and Pleasure is that according to such plot and design as shall be given you by our trusty and well-beloved Sir Christopher Wren, Knight, our surveyor-general of the place and site of the said observatory, you cause the same to be built and finished with all convenient speed.

So Wren, or rather, Wren's firm, was in charge of the design and construction of the observatory. This meant a significant role for Hooke the architect, as well as his input of scientific expertise, and the donation of some of his instruments and designs for others for the observatory. Wren appointed Hooke as site manager, 'to direct [the] Observatory in Greenwich Park for Sir J More', who was the King's representative. There is no detailed record of Hooke's contribution, but it clearly was substantial: on 28 July 1675 his diary records that he 'set out' the observatory (that is, marked the building lines out on the ground) in accordance with the architectural drawings, but we do not know if it was Wren, or Hooke, or both of them together, who made those drawings. Whoever was responsible, the building went up with impressive speed, and the observatory was essentially complete (at a cost of just over £520) by June 1676, when it was used to view an eclipse. The diary also records visits by Hooke to Greenwich in the company of Flamsteed and Halley, although we have no record of the exact date on which Hooke and Halley first met. It would be Halley, not Hooke, who played a major part in the scientific development of the Royal Greenwich Observatory, and who would become the second Astronomer Royal. It is time to fill in the background of this nineteen-year-old prodigy.

CHAPTER FIVE

FROM HACKNEY TO
THE HIGH SEAS

According to Samuel Pepys, 'Mr Hawley – May he not be said to have the most, if not to be the first Englishman (and possibly any other) that had so much, or (it might be said) any competent degree (meeting in them) of the science and practice (both) of navigation. And the inferences to be raised therefrom.' This passage is doubly revealing. First, thanks to the flexibility of spelling in seventeenth-century England, it shows us how to pronounce Edmond Halley's surname – for the first syllable to rhyme with 'trawl', or 'crawl'.* Second, it indicates the breadth of Halley's interests and skills. He is most widely known today as an astronomer, but here Pepys is praising his ability as a navigator, not just an expert in the theory of navigation, but as a practical seafaring explorer. There is much more to Edmond Halley than you might think, if you only know him because of the comet that now bears his name.

---

* The spelling we use – Edmond Halley – is the same as on his marriage certificate, and on his will.

We are particularly grateful to diarists such as Pepys, Hooke and Aubrey for the occasional glimpses they give us into Halley's life, because he kept no diary himself, and details of his personal life are scarce. We are not even absolutely sure when he was born, although the accepted date is 29 October 1656 (OS). Although Halley himself gave this date, doubts have arisen because although there is no surviving record of his birth, there is a record of the marriage of his parents, Edmond Halley senior and Anne Robinson, just seven weeks earlier than the accepted birth date. One suggestion is that the year of Halley's birth should be 1657, not 1656; a more likely possibility is that his parents had a civil marriage earlier, at the height of the Protectorate, and a church wedding some time afterwards. This is of no real importance, except that it demonstrates the difficulty of finding out even the basics about Halley's personal life.

At least we know something about his father, because Edmond Halley senior was a successful businessman who left his mark in the official records. His businesses involved the manufacture of soap and salting, an essential means of preserving food in those days, and he also owned property in London (that is, within the City itself) that brought in about £1,000 a year in income at the time young Edmond was born. As far as it is possible to make comparisons, that would be well over £100,000 today. As well as the properties that he rented out, Halley senior owned a house in Winchester Street, long since cleared to make way for railway lines running into Broad Street station, and a house at Haggerston, three miles to the north-east of St Paul's Cathedral. In 1656 this was countryside, part of the borough of Hackney. Edmond junior was born in the country house, but the family moved between the two residences. He had a sister, Katherine, who was born in 1658 and died as an infant, and a brother, Humphrey, whose birth date is not known, but who died in 1684. Anne Halley died in 1672, but the circumstances are not known. Unlike Hooke, Halley was not affected by the turmoil of the Protectorate and the Restoration of Charles II, which took place in 1660, when he was three.

A rare glimpse of the young Halley comes from Aubrey, who tells us that when the boy was about nine years old he was taught to write and do 'arithmetique' by an apprentice of his father. This might have been during the plague year of 1665, when presumably the family stayed in the comparative safety of their country home.

The following year, the fortunes of the Halley family were hit by the great fire, which destroyed a large part of Edmond senior's investment portfolio. But he weathered the storm, and when London was rebuilt he was able to charge higher rents for the improved buildings – an early link between Hooke and the Halley family. The salting business also prospered through the expansion of the navy under Charles II (and his able administrator Samuel Pepys), since salt meat was the staple diet in ships.

Very little is known about Halley's early education. He attended St Paul's School, but we are not sure when he started there. We do know that he was made School Captain in 1671, a year before Thomas Gale became headmaster, or High Master as he was officially titled. Gale, a distinguished scholar who had previously been in charge of Greek studies in Oxford, would play an important part in Halley's later career. He was an expert mathematician, and must have stimulated Halley's efforts in that direction, before the young man, encouraged by Gale, went up to Oxford in 1673, the year he was seventeen. If we are to believe Halley's contemporary Anthony Wood,* he arrived in Oxford with a glowing reputation:

He not only excelled in every branch of classical learning, but was particularly taken notice of for the extraordinary advances he made at the same time in the Mathematicks. In so much, that he seems not only to have acquired almost a masterly skill in both plane and spherical Trigonometry, but to be well acquainted with the science of Navigation, and to have made great progress in Astronomy before he was removed to Oxford.

---

* Anthony Wood, *Athenae Oxoniensis*, Oxford UP, 1721.

This should perhaps be taken with a pinch of salt, since it was written long after the event, when Halley was already famous. But it is certainly true that Halley arrived in Oxford not just with a reputation as an astronomer but also armed with astronomical telescopes and other equipment that would have been the envy of just about any professional astronomer of the day. His own later account, in the introduction to his *Catalogue of Southern Stars*, says that 'from my tenderest youth I gave myself over to the consideration of astronomy', while the study of science gave 'so much pleasure as is impossible to explain to anyone who has not experienced it.' And 'I soon realized that it is not from books one should try to advance this Art [astronomy], but that one ought to have recourse to instruments for measuring . . . ' He was lucky to have a father able and willing to provide him with the finest astronomical instruments, and clearly would not be the kind of dilettante Oxford undergraduate who enjoyed the high life, rarely attended lectures, and left without a degree. Although, as it happens, he did leave without a degree.

Halley entered Queen's College on 24 July 1673 as a commoner. This was, as the name suggests, the most common kind of undergraduate. 'Upper commoners' were the sons of nobility or wealthy gentlemen, and rarely took degrees. Commoners were the sons of gentlemen of a slightly lower status, and about half of them took their degrees with a view to becoming ordained. There were also subsizars (like Isaac Newton in Cambridge, and as Hooke had been, at least on paper) who earned their place by acting as servants for the wealthy students. Halley must have studied Latin and Greek, as well as mathematics, and on his own initiative he observed the heavens with his instruments, which included a telescope twenty-four feet long and a large quadrant. He learned geometry from John Wallis, one of the ablest mathematicians of the time and a founder member of the Royal Society, and astronomy from Edward Bernard, a former student of Wallis. Bernard had learned Arabic in order to study the Islamic texts that had preserved the astronomical knowledge of the Ancient Greeks, and Halley picked up enough for him to be able to study the Arabic texts himself.

While he was in Oxford, and also when at the Winchester Street house during vacations, Halley had been making astronomical observations in the company of Charles Bouchar, a slightly older student. Bouchar had been corresponding with the astronomer John Flamsteed, but left for Jamaica on family business early in 1675. So on 10 March that year Halley picked up the correspondence, writing to Flamsteed to explain why Bouchar had not replied to Flamsteed's latest letter, and describing his own observations. This was just a week after Flamsteed had been appointed as the first Astronomer Royal by Charles II. The most important content of the letter was that Halley had found that the positions of Jupiter and Saturn that he had observed differed significantly from the predicted positions published in tables based on calculations of their orbits. He asked if Flamsteed had 'observed anything of the like nature' and requested that, if so, 'I beg you would communicate it'.

Halley had also observed a lunar eclipse, this time from Winchester Street, on 1 January 1675. Once again, the timing of this event did not match published predictions, and once again he asks Flamsteed to confirm that he too noticed the discrepancy, And then, with great chutzpah, he says that he 'thinks fit to signifie' that he has found errors in some of the positions of stars recorded in the catalogue of Tycho Brahe, the great astronomer who had surveyed the northern skies in the 1590s. He sums up:

> These Sr. as a specimen of my Astronomical endeavours I send you, being ambitious of the honour of being known to you, of which if you shall deem me worthy I shall account my self exceedingly happy in the enjoyment of the acquaintance of so illustrious and deserving a person as yourself.

He was still only eighteen; Flamsteed himself not quite twenty-nine.

Flamsteed was suitably impressed with Halley's ability and enthusiasm, and in the summer of 1675 they made observations together from London, and were often seen chatting together in coffee houses. Flamsteed mentioned Halley in a letter to the

German astronomer Johannes Hevelius, and wrote in the *Philosophical Transactions* that 'Edmond Halley, a talented young man of Oxford, was present at these observations and assisted carefully with many of them.' As we have seen, Halley also accompanied Hooke and others on visits to Greenwich to plan the building of the Royal Observatory. Flamsteed moved to Greenwich in July 1675, and began observing from the site in September; Halley was back in Oxford by then, but continued to send Flamsteed his observations. With Flamsteed's encouragement, he produced three scientific papers in what turned out to be his last year in Oxford as an undergraduate.

The first was a theoretical paper, which demonstrates Halley's ability as a mathematician. He devised a geometrical technique to calculate the orbits of the planets using observations from the Earth, which, of course, is itself moving. This is by no means a trivial problem, and the paper helped to establish Halley's reputation. The second paper Halley was involved in was actually written by Flamsteed, and described their independent observations of dark spots that appeared on the surface of the Sun in July and August 1676. These spots were particularly noteworthy at the time, because sunspots were very rare in the second half of the seventeenth century; we now know that the Sun was in an unusually low state of activity at that time, which may have been linked to the cold of the so-called Little Ice Age that gripped Europe for decades. The third paper compared observations of an occultation of Mars by the Moon (when the Moon passes in front of Mars) made by Halley in Oxford, Hevelius in Danzig, and Flamsteed at Greenwich. In the same year, Halley devised, but did not publish, a technique to predict exactly where the Moon's shadow will pass over the surface of the Earth during a solar eclipse; Flamsteed later discussed Halley's idea in his book *Doctrine of the Sphere*, published in 1681.

By the autumn of 1676, Halley had completed three years as an undergraduate. In those days, it was usual (for those actually taking a degree) to spend four years in residence before graduating. But Halley had other ideas, which had been brewing in his mind for some time. One of Flamsteed's principal tasks as Astronomer Royal was to make a new survey of the heavens, improving on Tycho's survey not least because, unlike Tycho, he had telescopic sights to work with. This was all well and good, but what about the southern hemisphere? As Halley, and others, appreciated, it was high time for a survey of the southern skies, both out of scientific interest and in order to provide aid to navigation as ships ventured south of the equator. Halley was sure he was the right man for the job, and was too impatient to wait until he graduated to do it. Besides, someone else might pre-empt him if he delayed. After consultation with Flamsteed, he settled on the island of St Helena, the most southerly piece of land then held by the English, as the site for his observations. He was encouraged in this choice by what turned out to be misleading reports that the weather there was good and he would have clear skies for his observations.

As early as July 1676 Halley wrote to Oldenburg at the Royal, announcing this plan and presumably hoping for some sort of formal endorsement from the Society. He didn't succeed in getting an official endorsement, but Oldenburg joined Flamsteed in wiring a statement pointing out the value of the project, which Halley was able to use in an approach to the King for permission to go ahead. St Helena was governed by the East India Company, and one of its directors just happened to be Robert Boyle. The company had an original charter from Elizabeth I, reinforced by charters from Charles II, which essentially gave it a free hand to trade in the English interest, especially in India. It ran its affairs like a government, with its own ships and armed forces, governing territories like St Helena. But they were still subjects of the Crown, and had to take notice when they received a letter from the King 'recommending':

that Mr Edmund Hally a Student at Queen's College in Oxford, with a friend of his might have their passage on the first ship bound for St Helen's whether they are desirous to go & remayn for some times to make observations of the planets & starrs, for rectifying and finishing the celestial globe, being a place (he conceives) very fit and proper for that design; and that they may be received and entertained there, and have such assistance and countenance from the Compas [Company's] officers as they may stand in need of. On consideration whereof had, It is ordered that Mr Hally with his friend doe take their passage for St Helena on the Unity with their necessary provisions free of charge; and that a lre [letter] be written to the Governor & Council of the said Island to accommodate them with convenient lodging during their stay there, and afford them such assistance and countenance as may be for their encouragement, to proceed in so useful an undertaking.*

Halley's plans must have been well advanced, because that letter from the King is dated 4 October 1676, and the *Unity* sailed at the beginning of November, just after Halley's twentieth birthday. His friend, a Mr Clerke, sailed with him to assist with the observations, and Halley was provided with the best available astronomical instruments (in some ways, better than the ones Flamsteed had in Greenwich at the time), thanks to his father. His father also gave him an allowance of £300 a year, exactly three times the salary of the Astronomer Royal. The ship was owned by a family who lived near the Halleys in Winchester Street, which no doubt smoothed the arrangements.

The voyage itself played an important part in Halley's education. There is no hint of him ever having been to sea before, but as a passenger on a voyage that took three months to reach its destination (and on a similar ship for the voyage home), he had ample time to study the workings of such a vessel, and would have been particularly interested in the techniques used for navigation, which largely depended on astronomical observations. Indeed, it would be surprising if the captain had not taken advantage of having an experienced astronomer on board to consult

---

* Quoted in *The Observatory*, volume 51, 1928, and by Cook.

him about these observations. Halley would also have had an excellent opportunity to study the techniques of sailing itself – the way the ship was run, and the relationship between the officers and the ordinary sailors, as well as the mechanics of sail-setting and steering in different wind and weather conditions. We can only speculate, but he must have put this opportunity to good use, because, as we shall see, a decade later when he went to sea again he proved to be an accomplished seaman and navigator.

Halley and Clerke arrived at St Helena at the end of the southern hemisphere summer, and had to set up an observing site in rugged country high up on the slopes of the island's largest mountain, Diana Peak, before they could begin work (the site overlooks the location of the later site of Napoleon's tomb). When they began their astronomical work, they were plagued by bad weather, not just clouds, fog and rain but also by 'mighty winds', as Halley wrote in a letter to Sir Jonas Moore. Much later, in a paper published in the *Philosophical Transactions* in 1691, Halley recalled:

> In the Night time, on the tops of the Hills about 800 yards above the Sea, there was so strange a condensation or rather precipitation of the Vapours, that it was a great Impediment to my Celestial Observations; for in the clear Sky the Dew would fall so fast as to cover, each half quarter of an hour, my Glass with little drops, so that I was necessitated to wipe them so often, and my Paper on which I wrote my Observations would immediately be so wet with the Dew, that it would not bear Ink.

They were also hampered by the uncooperative attitude of the governor of the island, Gregory Field; this does not seem to have been anything personal, since Field was so unpleasant and uncooperative to everyone that he was sacked by the Company in February 1678, too late to do any good for Halley. In spite of the difficulties, Halley and his assistant were able to measure the positions of 341 stars relative to the positions of those stars in

Tycho's catalogue which are visible from St Helena. Although Halley realised that Tycho's catalogue would need revision as better observing instruments were developed, by locking his catalogue in to Tycho's he ensured that, when Tycho's positions were improved, his positions would still be accurate relative to the improved northern catalogue. His was the first major survey of this kind using telescopic sights, in either hemisphere, and remained the best guide to the southern skies well into the following century.

But the catalogue was only one of the fruits of Halley's stay on St Helena. He took with him a pendulum clock, in pieces to be assembled on site, to help with timing his astronomical observations. He found that in order for the clock to keep correct time (in effect, to make sure that it ticked off exactly twenty-four hours in each day, from noon to noon by the Sun) the pendulum of the clock had to be shorter than when the same clock was in England. We now know that this is because the Earth bulges at the equator as a result of its rotation, so the effective gravitational pull felt at low latitudes is less than that felt at high latitudes. Robert Hooke was the first person to appreciate this, and, as we explain later, he used Halley's measurements as evidence in support of the idea of a terrestrial equatorial bulge.

But the observations from St Helena which made the most long-standing impact on astronomy, and opened up a way to measure distances across the Solar System, were made on 28 October 1677, when the planet Mercury, as seen from Earth, passed across the face of the Sun. Such events, known as transits, are rare but predictable, and this particular transit of Mercury was one of the main reasons why Halley wanted to be on the island that year. Because observers in different parts of the world are viewing (in this case) Mercury from slightly different angles, they will see it cross the edge (limb) of the Sun at slightly different times. This is a simple example of parallax. By timing the exact moments when Mercury crossed the Sun's limb both going in (ingress) and out (egress) and comparing the interval between these measurements with similar measurements made in Europe, it would in theory be possible to use simple geometry and the laws of planetary orbits discovered by Johannes Kepler to work out the distance of the Earth from the Sun, in terms of multiples of the Earth's radius.* The idea had been known for some time, but Halley was the first to put it into practice. In spite of the terrible observing conditions he was able to write to Moore that:

> I have notwithstanding had the opportunity of observing the ingress and egress of Mercury on the Sun,† which compared with the like Observation made in England, will give a demonstration of the Sun's Parallax, which hitherto was never proved [that is, measured] . . .

Unfortunately England was covered by cloud at the critical time, and French astronomers at Avignon were only able to observe the egress, when Mercury moved out of the line of sight to the Sun, not the ingress. So it was only possible to get a rough estimate for the Sun–Earth distance, which later turned out to be only about a fifth of the true value. Halley realised, however, that a transit of Venus would give a much better opportunity to make this measurement, a realisation which would lead to one of his two great posthumous successes (see Chapter Eleven).

Halley and Clerke must have left St Helena only a few weeks after observing the transit, probably at the end of February, because in his diary entry for 30 May 1678 Hooke writes 'met Hally from St Helena with S[ir]. J. Moore & Colwell at toothes.' Toothes was one of the coffee shops frequented by Hooke and his friends. But over the following weeks Halley was too busy to spend much time in coffee houses. He immediately set to work on preparing his star catalogue, in which he included an account of how he had become interested in astronomy, descriptions of

---

* The timing comparison depended on knowing the longitude of St Helena, but this was straightforward to determine from astronomical observations made from the stable land surface rather than from the heaving deck of a ship.

† In the letter he uses as the usual shorthand the astrological symbols for Mercury and the Sun.

other objects he had observed from St Helena, such as star clusters and the Magellanic Clouds, as well as discussions of why the predicted positions of Jupiter and Saturn were incorrect in the published tables, and an analysis of the orbital motion of the Moon. He gave the resulting treatise an unwieldy title which began *A Catalogue of the Southern Stars, or a supplement to the Catalogue of Tycho* . . . and ran on for a total of more than a hundred words. Hardly surprisingly it is usually known simply as *A Catalogue of the Southern Stars*.

Halley's catalogue was published (in Latin) early in November 1678. Hooke described the work to a meeting of the Royal Society, and at the end of the month the man Flamsteed now referred to as 'our southern Tycho' was elected as a Fellow. This honour was soon followed, on 3 December, by another, at the express wish (after a little prodding) of the King.

Halley had made an overt attempt to butter the King up by inventing a 'new' southern constellation, which he called *Robur Carolinium* ('Charles' Oak') in recognition of the oak tree in which Charles had hidden after the Battle of Worcester. The name did not stick, and no such constellation is now recognised, but no doubt Charles was pleased. There had clearly also been some behind-the-scenes machinations concerning Halley's academic status, because on 12 November the Provost of Queen's College had written to the Secretary of State, Sir Joseph Williamson, picking up on earlier communications, now lost. 'I have spoken with Mr Vice-Chancellor about procuring a degree for Edmond Halley,' he wrote. 'If you will procure the King's letter, it will be both effectual and not unpleasing to the University. He is now of almost 6 years standing and less than a master's degree cannot be conferred on him.' The reason the university needed a letter from the King was that there were (and are) strict rules about the amount of time undergraduates had to reside at the university, which clearly Halley had not kept; nor had he taken any examinations. But the King was happy to oblige. This was not entirely unprecedented; other people had been awarded degrees at the King's behest. But on this occasion it is worth quoting his letter, dated 18 November, at length, because as a result of it Halley received the first degree ever awarded specifically for research:

. . . having received a good account of the proficiency in Learning of our Trusty Wellbeloved Edmund Hally of Queens College in that our University especially as to the Mathematick and Astronomy, Whereof he has (as We are informed) gotten a good Testimony by the Observation he has made during his abode in the Island of St. Helena; We have thought fit for his Encouragement hereby to recommend him effectually to you for his Degree of master of Arts; Willing and Requiring you forthwith upon the receipt hereof (all Dispensations necessary being first granted) to admit him the sd. Edmund Hally to the said Degree of Master of Arts without any Condicion of performing any previous or subsequent Exercises for the same any Statute or Statutes of that Our University to the contrary in any Wise notwithstanding . . .

The earliest opportunity the University had to confer the degree was at a Congregation held on 3 December, by which time Halley was already a Fellow of the Royal Society. He had arrived with a bang on the scientific scene, displaying a multiplicity of skills which has been neatly summed up by Alan Cook:

In his first and early major campaign Halley showed an outstanding ability to plan and persuade, to organise and to carry through his project. He showed theoretical insight into fundamental matters of astrometry, and acquired more practical experience as an observing astronomer with up-to-date instruments than anyone else of his day. He had accomplished the three things he set out to do, to observe in the south, to produce the first catalogue after Tycho, and to use modern instruments in a major systematic campaign.

This was clearly a young man – only just twenty-two in November 1678 – who was going places. And the next place he went, this time on semi-official Royal Society business, was Danzig; that business also involved Hooke.

Johannes Höwelke was an astronomer of the old school, born in 1611, who, as was the custom at the time, particularly on the continent, had Latinised his surname and was known to all as Hevelius. He came from a wealthy family of brewers, and was able to indulge his passion for astronomy by building an observatory in Danzig (spread across the roofs of several houses), which was the best in Europe until the construction of the Greenwich and Paris Observatories. In 1662, Hevelius' first wife died, and the following year he married a girl of sixteen, Elizabetha. The daughter of a merchant, she soon proved to have a good business head herself, more or less running the brewing business and helping her husband with his astronomical observations. His work was known and admired across Europe, and he was elected as a Fellow of the Royal Society in 1665. But doubts about his work began to emerge in the 1670s.

As an astronomer of the old school, Hevelius made all his observations using open sights. In 1673, he published a book (part one of an opus called *Machina coelestis*) in which some of these observations were reported. After the publication of this book, Flamsteed and Hooke were among the astronomers who criticised the old-fashioned technique used by Hevelius. Flamsteed wrote in the *Philosophical Transactions* that:

> We have heard that the celebrated Johannes Hevelius has indeed undertaken the restitution of the fixed stars, yet seeing he is reputed to use sights without glasses, it is doubtful if we shall obtain from him much more correct places than Tycho left us.

Hooke, always an innovator and never afraid to speak his mind, pointed out the deficiencies of open sights and the advantages of telescopic sights in his second Cutlerian Lecture in 1674. The lengthy title of the published lecture begins 'Animadversions on the first part of the Machina Coelestis', which leaves the reader in no doubt where Hooke is coming from. Hooke goes into great detail in describing and discussing both the instruments used by Hevelius and his own instruments, and referring to previous criticism of the telescopic sights made by Hevelius in correspondence with Hooke. His exasperation is clear:

> [I] am sorry I have been forced to say so much in vindication of Telescopical Sights; and that in the doing thereof, I have been necessitated to take notice of the imperfections, that are the inseparable concomitants of Instruments made with Common [open] Sights. Nor should I have published these my thoughts, had I not found them so highly decryed by a person of so great Authority, fearing that hereby other Observators might have been deterr'd from making any use of them, and so the further progress of Astronomy might have been hindred.

Hooke is reasonableness itself in summing up the situation. He refers to Hevelius as an 'excellent Person' who in his astronomical work 'seems not to have spared either for labour and vigilancy, or for any cost and charges that might check his purpose, for which he hath highly merited the esteem of all such as are lovers of that Science'. But, says Hooke:

> if he had prosecuted that way of improving Astronomical instruments, which I long since communicated to him, I am of opinion he would have done himself and the learned World a much greater piece of service . . . by publishing a Catalogue ten times more accurate.

Hevelius was unconvinced, and the matter rumbled on. Perhaps 'rumbled' is too mild a term. Hevelius developed a great antipathy towards Hooke, as shown by a letter he wrote to Flamsteed on 14 June 1676:

To be honest, I repeatedly have serious doubts as to whether, with his telescopic lenses and polemoscope mirrors, or with any other devices whatsoever, that man (I repeat, *that* man)* will be able to produce results more definitive or consistently more complete than ours.

Without modifying his views on Hooke, early in 1677, Hevelius wrote in more friendly terms to Flamsteed, saying 'let each allow the other to help on Astronomy at his own risk and by his own methods'.† He also wrote to Oldenburg with details of his observations. But the matter could not be left entirely for each astronomer to work in his own way and with his own instruments. That was all very well in Tycho's day, but by the late 1670s astronomy was becoming an international endeavour that needed recognised standards. Apart from the work done by Hevelius, Flamsteed was carrying out a detailed survey of the northern skies using telescopic sights, and Halley had already carried out the first survey of the southern skies. All of these observations (and those from Paris and other places) had to be matched up with one another and combined to chart the heavens. It was crucial to know how far Hevelius' published results could be trusted. Sure that he would be vindicated, Hevelius invited the Royal Society to send an astronomer, armed with telescopic instruments, to make observations from Danzig and compare these with observations made by the same observer using the open sights instruments at the Danzig observatory. There was an obvious choice. Halley was already an experienced observer, had no ties, his own portable instruments, and plenty of money to finance his own travel without costing the Royal a penny. He was trusted by Flamsteed, and had sent Hevelius an early copy of the *Catalogue of the Southern Stars*, with a letter in which he said that he would be happy to recalculate his star positions relative to the northern survey then being carried out by Hevelius, instead

---

* Emphasis in the original.
† Translation by Ronan.

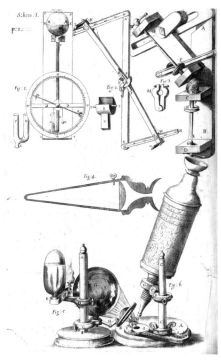

above: View towards Rocken End (Isle of Wight) from Whale Chine with, in the foreground, gravel and blown sand above the Cretaceous, ferruginous sands. © Ian West

right: Some of Hooke's equipment, including (bottom right) his microscope. © Royal Institution of Great Britain/ Science Photo Library

below: The title page of Hooke's great book. © The Royal Society

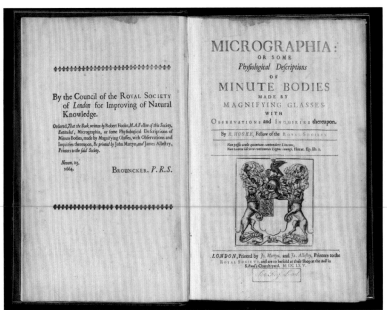

By the Council of the ROYAL SOCIETY of *London* for Improving of Natural Knowledge.

Ordered, That the Book, written by Robert Hooke, M.A Fellow of this Society, Entituled, Micrographia, or some Physiological Description of Minute Bodies, made by Magnifying Glasses, with Observations and Inquiries thereupon, Be printed by John Martyn, and James Allestry, Printers to the said Society.

Novem. 23, 1664.

BROUNCKER. *P. R. S.*

MICROGRAPHIA:
OR SOME
Physiological Descriptions
OF
MINUTE BODIES
MADE BY
MAGNIFYING GLASSES.
WITH
OBSERVATIONS and INQUIRIES thereupon.

By R. HOOKE, Fellow of the ROYAL SOCIETY.

Non possis oculo quantum contendere Linceus,
Non tamen idcirco contemnas Lippus inungi. Horat. Ep. lib. 1.

LONDON, Printed by *Jo. Martyn*, and *Ja. Allestry*, Printers to the ROYAL SOCIETY, and are to be sold at their Shop at the *Bell* in S. Paul's Church-yard. M. DC. LX V.

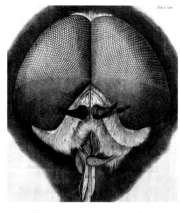

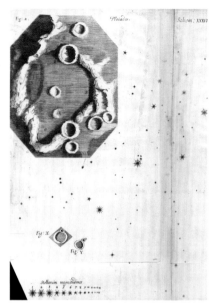

above: The eyes and head of a fly, from *Micrographia.* © Natural History Museum, London/Science Photo Library

right: Hooke's drawing combining a star map with a detail of the Moon showing craters. © The Royal Society

above: A view of the dome of St Paul's, designed by Wren using a technique devised by Hooke.
© Historical Images Archive/Alamy

left: One of Hooke's churches, St Benet Paul's Wharf. © Michael Grant/Alamy

right: An engraving of the Monument, London from a book printed around 1850. © Stephen Dorey/ Bygone Images/Alamy

below: Ramsbury Manor, Wiltshire, designed by Hooke. © Chronicle/Alamy

below: Coloured engraving of the Royal Observatory at Greenwich at the time of Flamsteed. © Science Photo Library

above: View of the Frost Fair on the Thames looking downstream towards London Bridge during the Great Frost of 1683–4. © Sheila Terry/ Science Photo Library

below: Engraving of Halley's diving bell of 1716. © Sheila Terry/Science Photo Library

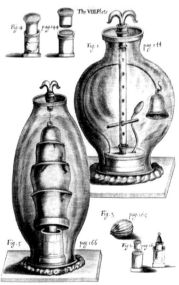

above: Apparatus used by Hooke and Boyle to demonstrate that the sound of a ringing bell cannot travel through a vacuum. © Royal Astronomical Society/Science Photo Library

right: Johannes Hevelius at left, carrying out astronomical observations with his second wife Elisabetha Koopman Hevelius. © Royal Astronomical Society/Science Photo Library

below: Hevelius's plan for a tower observatory, showing various telescopes, including his 150-foot refracting telescope. © World History Archive/Alamy

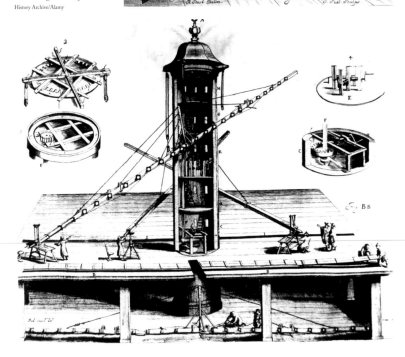

left: A 'Pink', similar to the ship commanded by Halley. © Lordprice Collection/Alamy

The Pink

below: Halley's wind chart (see text), published in 1686.
© Royal Astronomical Society/ Science Photo Library

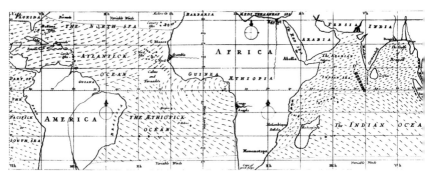

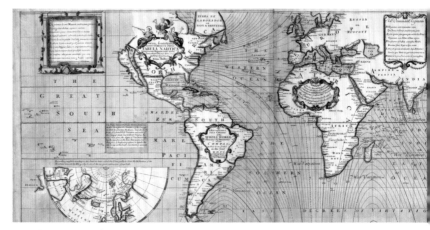

above: Halley's chart of magnetic compass readings, published in 1700.
© Stephen A. Schwarzman Building/The Lionel Pincus and Princess Firyal Map Division/New York Public Library/Science Photo Library

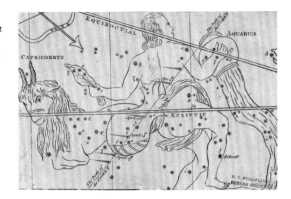

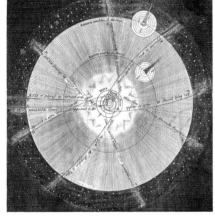

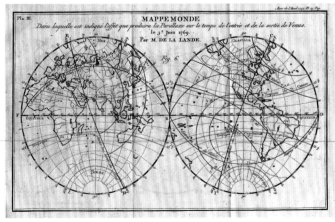

above: Map prepared by Halley showing where the transit of Venus
due in 1761 could be observed.

above: Halley's Comet, photographed from Peru on 21 April 1910. © Harvard College Observatory/Science Photo Library

left: A map of England drawn by Halley showing the passage of the shadow of the Moon during the total solar eclipse of 22 April 1715. The dark circle gives the size of the Moon's shadow at a given time during the eclipse; the diagonal stripe shows locations from which totality was observed.
© Science Photo Library

below: A graph showing the sunspot group number as measured over the past 400 years. The Maunder Minimum, between 1645 and 1715, when sunspots were scarce and the winters harsh, is clearly visible.
© WDC-SILSO

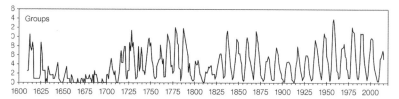

of relying on Tycho's catalogue. In April 1679, armed with a scarcely necessary letter of introduction from the Royal, Halley set sail for Danzig on what was technically a private visit, although Hevelius seems to have regarded him as an official envoy of the Society. He arrived on 26 May, and the two astronomers made observations together, sometimes with others as well, every clear night for the next two months.

It turned out that the measurements made by Hevelius were more accurate than Halley had anticipated, largely because Hevelius used two observers to make the measurement, one lining up a sight on one star while the other moved a second sight to another star to measure the separation of the two stars, from the angle between the two sights. This was better than having a single observer use one sight that was moved from the first star (after noting its position) to the second star. Impressed, Halley wrote to Flamsteed on 7 June with a description of the instruments and the observatory, and then discussed the measurements of star positions made by different observers. 'I assure you I was surpriz'd to see so near an agreement in them, and had I not seen, I scarcely could have credited the Relation of any; Verily I have seen the same distance repeated several times.' The measurements also agreed with those obtained using Halley's own instruments, with telescopic sights. But this was hardly a fair comparison, because the portable instruments could not be expected to be as accurate as permanent, full-size instruments like those Flamsteed was now using in Greenwich. As Halley must have been aware, but did not point out, this implied that Flamsteed's instruments were more accurate than those of Hevelius.

Halley and Hevelius clearly got on very well with one another during the weeks they were observing together, and in mid-July, shortly before Halley left Danzig, Hevelius asked him to provide what amounted to a testimonial to the older man's observing technique. The document Halley provided makes fascinating reading, and suggests that Halley was a natural diplomat, tactful to the point of being disingenuous:

I have seen with my own eyes not one or two, but several observations of the fixed stars made with your large brass sextant by different observers, and some of them by myself, which even when repeated did most accurately and almost incredibly agree . . .

Hardly surprisingly, Hevelius referred to Halley as 'a very pleasant guest, a most honest and sincere lover of the truth'. But look carefully at what Halley actually said, both in the testimonial and in the letter to Flamsteed. Nowhere does he provide a comparison with measurements made using other instruments, and nowhere does he say that the measurements made by Hevelius are as accurate as Hevelius claimed. He says that repeated measurements made with the same instruments gave the same answer. Flamsteed and Halley were pioneers of the application of statistical methods in astronomy to determine what we now call systematic errors – that is, errors that are built into particular instruments, so that although they always give the *same* answer it is not necessarily the *right* answer. And it seems Halley said one thing for public consumption and another in private, since in his diary entry for 14 August 1679 Hooke wrote: 'Halley returned this day from Dantzick (Hevelius rods in pisse).' The tact, if you like to call it that, would serve Halley and the Royal Society well when the time came to cajole Newton into publishing his masterwork. In spite of his efforts, however, the controversy concerning Hevelius continued to rumble on until it died with him in 1687.

A couple of years before his death, Hevelius published a book, *Annus Climactericus*, which was riddled with errors about Halley's work. He said that Halley had been sent to St Helena by the Royal at Hevelius' suggestion, when in fact he had gone on his own initiative; he described Halley's instruments incorrectly; and generally downplayed Halley's work. On 26 March 1686, Halley wrote to a colleague in Dublin, William Molyneux, that:

the Controversy between Mr Hevelius and Mr Hook, as you very well observe does, as Hevelius manages the matter, affect all those observers that use Telescopic sights, and myself in particular, and it is our common concern to vindicate the truth from the aspersions of an old peevish gentleman, who would not have believed it possible to do better than he has done, and for my own part I find myself obliged to vindicate my observations made at St Helena and to rectify some mistakes, whether wilful or not I cannot say; first he sais I was sent by R.S. to St Helena at his request to observe the Southern Starrs (pag. 14) whereas it is very well known to all our Astronomers that at my own motion and charge I undertook that voiage above two years, before I had the honour of being a member of the R. S. . . .

In a second letter to Molyneux, dated 27 May 1686, he hints at further revelations:

As to Mr Hevelius we heare as yet no farther from him, and I am very unwilling to let my indignation loose upon him, but will unless I see some publick notice taken elsewhere, let it sleep until after his death if I chance to outlive him, for I would not hasten his departure by exposing him and his observations as I could do and as I truly think he deserves I should.

In fact, Halley never did 'expose' Hevelius and his observations; there was no need, since the death of Hevelius brought an end to the controversy as there was nobody left to champion the cause, such as it was, of open sights.

Back in 1679, however, Halley seems to have been at something of a loose end after his trip to Danzig. He spent some of his time in Oxford and part of it in London, where we get glimpses of him from mentions in Hooke's diary and other sources. We can pick up his trail properly again in December 1680, when he left for the Grand Tour through France and Italy that was part of the usual upbringing of a young gentleman. He travelled with a friend, Robert Nelson, who was the same age as him (twenty-three at the start of the trip); Nelson was the son of a wealthy merchant and a Fellow of the Royal Society, although one of the gentlemen Fellows with a dilettante interest in science, not an active scientist.

It was just before the two gentlemen set off on their tour that Halley had his first encounter with a comet. A bright comet visible to the naked eye from London was a major talking point in the city in November 1680, and naturally Halley, Hooke and others made observations of the way it moved against the background of stars as it closed in on the Sun. It was soon lost to sight in the glare of the Sun, but Halley saw it again, now moving away from the Sun, after he had crossed the English Channel (a very rough and unpleasant crossing) and was on the way to Paris. In one of his notebooks, Isaac Newton recorded that 'Hally has told me that on his journey to Paris on Dec 8 old style he saw the tail of the comet rising perpendicularly from the horizon', a form of words which suggests that soon after his return to England (early in 1682) Halley met Newton and discussed the comet with him. This would have been the first time they met each other.

Already, Halley's version of the Grand Tour was deviating from that of most young men. The usual focus of attention was the art, culture and historical sights of Europe. But Halley's tour was as much a scientific expedition as anything else, and in Paris he lost no time in visiting Giovanni Cassini,* the head of the Paris Observatory, to compare notes and make more observations. The large, bright comet was as much a sensation in Paris as it had been in London, with many people seeing it as a portent of disaster, a supernatural phenomenon.

So little was known about comets at the time that there was some doubt about whether what had been seen was one comet or two, one moving towards the Sun and one moving away from it. Flamsteed was one of those who subscribed to the view that it was indeed one object, but even Flamsteed did not realise that it was following an orbit around the Sun. He envisaged that some sort of magnetic effect had first attracted the comet

---

* Although Italian, because Cassini worked in France he is also known as Jean Cassini.

towards the Sun and then repelled it. On 15 January 1681 Halley wrote to Hooke that:

> The general talk of the virtuosi here is about the Comet, which now appears, but the cloudy weather has permitted him to be but very seldom observed. Whatever shall be made publick about him here, I shall take care to send you, and I hope when you shall please to write to me you will do me the favour to let me know what has been observed in England.

Halley moved on from Paris to Saumur, from where he wrote to Hooke again in May, including news of the latest ideas about the comet:

> Monsieur Cassini did me the favour to give me his books of ye Comett Just as I was goeing out of towne; he, besides the Observations thereof, wch. he made till the 18 of March new stile, has given a theory of its Motion wch. is, that this Comet was the same with that appeared to Tycho Anno 1577, that it performs its revolution in a great Circle including the earth . . .

Halley says that Hooke might find it hard to accept all this, but that it is certainly an idea worth thinking about. As it happens, Cassini was mistaken in thinking that the 'new' comet was a return of the one seen by Tycho Brake in 1577, but it does seem that he planted an idea in Halley's mind.

In the same letter, Halley describes to Hooke a comparison he has made of the cities of London and Paris, which would have been of great interest to Hooke the architect and city planner. By carefully pacing out the dimensions of Paris, Halley has found that it is 'not soe great a Continuum of houses as London', but 'by reason of theire living many in a house' it has a larger population. This is born out by the evidence of burials and christenings: in 1680, he points out, 24,411 people were buried in Paris, 'whereas at London 20000 is reckoned a high bill, and the Christenings farr exceed ours, haveing been almost 19000, when we have ordinarily 12 or 13000'. He then goes on to mention an insight, which has just come to him:

> halfe as many were married as were borne; and not more; it will from hence follow, supposing it alwaies the same, that one half of mankinde dies unmarried, and that it is necessary for each married Couple to have 4 Children one with another to keep mankind at stand. this Notion Occurred whilst I was writeing . . .

The notion that Halley tosses away in a letter – that populations produce more offspring than the number required to reach maturity and reproduce in their turn – was one of the key insights that two centuries later would lead Charles Darwin and Alfred Russel Wallace independently to the idea of evolution by natural selection. Halley never followed that path, but he did return to the study of population statistics, and later wrote a paper that laid the foundations of the actuarial approach to life insurance.

From Saumur, Halley and Nelson followed a leisurely route to Rome, arriving in October, and planned to spend a few months touring Italy; he certainly must have intended to visit Venice. In August, Halley had observed a lunar eclipse from Avignon, providing data that Flamsteed, who had observed the eclipse from Greenwich, was able to use to calculate the longitude of Avignon relative to Greenwich. Somewhere along the way, Halley also picked up some astronomical observations made in India, which he passed on to Hooke as they could be used to determine the longitude of the observing site in Ballasore. The exiled Queen Christina of Sweden was living in Rome at the time, and was sufficiently interested in astronomy that she had offered a prize for anyone who could successfully predict the return of the comet. The prize was never won, although several astronomers, including Cassini and Hevelius (but not Halley), tried for it. The Queen would have been eager to have Halley's first-hand account of the observations from London and Paris, and of the latest ideas about comets from the virtuosi. It seems highly likely that he met her, although there is no record of the meeting.

But then things took an unexpected turn – actually, two unexpected turns. In late November, less than a year after setting out on his travels, Halley was called back to England on family business, but it cannot have been too urgent since he travelled via Genoa, Paris and Holland, instead of by the most direct route. Nelson did not return with him; he fell in love with a widow, two years older than him, who had a twelve-year-old son by her first husband. They married in 1682; she was the second daughter of the first Earl of Berkeley, whose home at Durdans had provided a safe base for Hooke during the plague year.

Just why Halley had to cut short his Grand Tour is not clear. A plausible guess is that it had something to do with the second marriage of his father, who had been a widower since 1672. The exact date of the marriage is not known, but Halley was back in London on 24 January 1682, as noted by both Hooke and John Aubrey, who says: 'He hath contracted an acquaintance and friendship with all the eminent mathematicians in France and Italie'. Halley then goes off the radar until, out of the blue, we find that on 20 April 1682 he too got married, to Mary Tooke, the daughter of a prominent lawyer. The brief biography in MacPike describes her as 'an agreeable young Gentlewoman; and a Person of real merit'. We know nothing about the relationship of the couple before the wedding, and precious little afterwards. Although they were together for more than fifty years, the only surviving records concerning Mary are those of the birth of her children: Margaret, born in 1685, Katherine, born in 1688, and Edmond, born in 1698. Possibly there were other children who died in infancy. None of the three children we know about had children of their own.

These changes in Halley's life did not take him out of the mainstream of science in England. The couple lived in Islington, then a village just outside London, where Halley set up an observatory and continued to make astronomical observations, and from where he could easily make the short journey to attend meetings of the Royal Society. The key project that he embarked upon was a series of observations of the Moon, intended to cover the entire cycle, some eighteen years long, of its apparent movement against the background stars, with a view to producing tables that mariners could use to find their local time, and hence the longitude, by observing the position of the Moon. In August and September 1682 another bright comet was visible, and was monitored by Halley, along with his other observations, but more notably by Robert Hooke. This was not as sensational as the comet of 1680–81, but played an important part in the development of Hooke's ideas about gravity, as we discuss in Chapter Seven. It also later became significant in Halley's own investigation of the heavens. Halley published a couple of solid but unspectacular papers on astronomical topics in 1683, and measured the orbital period of the moon Titan (itself only discovered in 1655 by Huygens) around Saturn, coming up with a figure of fifteen days, twenty-two hours, forty-one minutes, close to the modern value. He appeared to have settled into a comfortable way of life, combining both domesticity and science, when his world was turned upside down by the sudden death of his father, in dramatic and unexplained circumstances, on 5 March 1684.

This was just two months after Halley, Hooke and Wren had debated the nature of gravity, and the puzzle of proving that it obeyed an inverse square law of attraction, after a meeting of the Royal Society. When family business related to the death of his father meant that Halley had to visit Alconbury, near Huntingdon, later in 1684, he took the opportunity to visit Newton in Cambridge to discuss these ideas with him, with far-reaching results. His father's death, and Newton's response to that visit, changed Halley's life dramatically; we shall pick up the threads in Chapter Eight, after giving Hooke's perspective on the story of gravity.

CHAPTER SIX

OF SPRING AND
SECRETARYSHIP

We left Hooke in 1677, after the hatchet-burying exchange of letters with Newton concerning their priority dispute about light. As Newton retreated once again from the scene, things soon got better still for Hooke with the death of his nemesis, Oldenburg. Among other things, this gave Hooke the opportunity to publish the one piece of work for which he always received credit and was never overshadowed, without having to worry about Oldenburg's 'snares'.

As we have seen, in the early 1670s Hooke was occupied with many projects, and the Royal Society was not top of his list of priorities. The Royal itself had declined from its early enthusiasm, and become more of a kind of gentlemen's debating club, with less real science being done, partly because Hooke was not so active. But he was still involved in the dispute with Hevelius, the Gregorian telescope, experiments with magnetic 'lodestones', and in particular watchmaking, alongside his surveying and architectural work. The watchmaking is particularly important to our story, because it was closely linked with his study of springs. Although Hooke never got a patent for his ideas, in the mid-1670s and later he was working closely with the master watchmaker Thomas Tompion, and later also with John Bennett, who produced a succession of spring-driven timepieces to Hooke's designs, which Hooke sold. The watches became widely used in Hooke's circle. It was through this work that he discovered his law of springs. We don't know when he discovered the law, but we do know when he first revealed it. On 2 September 1675 he wrote in his diary that he had told Tompion that 'all springs at liberty bending equal spaces by equal increases of weight'. A day later, he made a spring scale 'to shew the King', in which a vertical coiled spring was stretched by adding weights to the bottom, moving a pointer down a scale to indicate the weight. Here again is an example of the overlapping of Hooke's many interests; such a scale clearly had practical uses in, for example, building work. But with Oldenburg still Secretary of the Royal, it was not yet time for Hooke to present formally all his ideas about springs.

One of the ideas which Hooke described to Boyle (in April 1677) but never formally presented to the semi-moribund Royal Society was a 'wheelhorse', which seems to have been a proposal for a two-wheeled vehicle propelled by the feet – the kind of pedal-less bicycle, or velocipede, that was popular in the early nineteenth century. Such a vehicle would have appealed to Hooke, who hated horse-riding, but it seems he never built one. At the end of April, he observed a comet, which encouraged him to prepare a Cutler lecture on the subject of comets, which was later published.* This seems to have revived his interest in the way planets move around the Sun, which he discussed with Wren.

They knew, of course, about Johannes Kepler's laws of plane-tary motion. These empirical laws, based on the observations made by Tycho Brahe, encapsulate three facts. First, planets move in elliptical orbits, with the Sun at one focus of the ellipse. Second,

---

* The Cutler and Gresham lectures were themselves pretty moribund at this time, with few, if any, people attending them, but Hooke still carried them out in accordance with the terms of his contract.

planets move faster when closer to the Sun, and slower when further away, in such a way that an imaginary line linking the planet and the Sun traces out equal areas in equal times, at any part of the orbit. Third, the square of each planet's 'year' is proportional to the cube of its average distance from the Sun. It is actually possible to work out the inverse square law of gravity from Kepler's laws, as Newton later showed.* But Hooke and Wren did not do this. What Hooke did realise was that these laws imply that the force of the Sun's attraction acts more powerfully when the planet is closer to the Sun, and is weaker when the planet is further away.

But at that time this was still almost an aside from his architectural work. It was in March 1677 that Hooke presented his design for what became the Pepysian Library at Magdalene College, Cambridge, although he was not involved in the construction. And on a personal note, it was in August that year that Grace departed for her fateful visit to the Isle of Wight, a month after Hooke had taken on another assistant (and lodger), Thomas Crawley. The month after Grace departed, two deaths shook up Hooke's world. On 12 September, Tom Giles died of smallpox, shaking up Hooke's domestic life. But on 5 September, Oldenburg had died, probably of malaria, shaking up Hooke's scientific life and leading to a revival of the Royal Society.

On 13 September, which was also the day of Tom's funeral, Hooke took over the Secretaryship on a caretaker basis, until a permanent replacement for Oldenburg could be appointed. The Secretary was the key person in the Royal, essentially responsible for running the Society, keeping records, and communicating with scientists across Europe. Hooke, along with Boyle, Wren, Aubrey and others, realised that this was an opportunity to restore the Royal to what it had been intended to be. Some of the founders had died, others had got old and lost interest, but there was a rising generation ready to take over. After much lobbying and behind-the-scenes manipulating, on 30 November 1677 the

---

* 'Later' being the key word.

Fellows elected a new President, Sir Joseph Williamson (the Secretary of State), a new Treasurer, Abraham Hill (one of Hooke's camp), a new Vice-President, Christopher Wren, and joint Secretaries, Hooke and Nehemiah Grew (another curator of experiments). Hooke and Grew were also elected to the Council, the governing body of the Royal. Significant though this coup was, both for Hooke and for the Royal, in scientific terms it was accompanied, coincidentally, by an equally significant development in microscopy, with Hooke revitalised and throwing himself wholeheartedly back into scientific work.

Earlier in the year, news had come from the Netherlands of Anton van Leeuwenhoek's discovery of tiny living organisms wriggling about in droplets of water. Leeuwenhoek was very secretive about the methods he had used to observe these creatures, so that when Grew, in the spirit of *nullius in verba*, had tried to replicate the observations he had had no success. But on 10 November, using an improved microscope and the trick of trapping water in very thin tubes of glass, which themselves had a magnifying effect, Hooke saw in rainwater that had been standing for a week 'great numbers of exceedingly small animals swimming to and fro' and commented 'nor could I indeed imagine that nature had afforded instances of so exceedingly minute animal productions'. On 15 November, he showed these microorganisms at the Royal.

Hooke followed this up by making his own single-lens microscopes (essentially very powerful magnifying glasses) of the kind Leeuwenhoek had hinted that he used. These are tiny droplets of glass, fixed in a hole in a thin metal plate, which are extremely difficult to make and equally difficult to use, involving such close focusing with the eye that the operator's vision is likely to be damaged by long use (as, indeed, Hooke's seems to have been). Significantly, and unlike Leeuwenhoek, Hooke was eager to pass on the details not only of what he had discovered but how he had discovered it. In an essay, *Microscopium*, published in April 1678, he wrote:

The manner how the said Mr. Leeuwenhoek doth make these discoveries, he doth as yet not think fit to impart, for reasons best known to himself; and therefore I am not able to acquaint you with what it is: but as to the ways I have made use of, I here freely discover that all such persons as have a desire to make any enquiries into Nature this way, may be the better inabled to do so.

Although he was generally so open about his discoveries, there was one secret that Hooke had, until now, hugged to himself, for reasons that were practical and understandable – his theory of springs. But with Oldenburg dead and the prospects of a patent for his watches remote, the time had now come for him to release that secret into the world. And it was far more than the seeming simplicity of 'Hooke's law' might lead you to think.

A letter written by Hooke in January 1678, to Martin Lister, a botanist based in York, sums up the optimism with which he started the new year. The changes at the Royal, he said, had 'very much revived us and put a new spirit in all our proceeding which I perswade myself will not only be beneficial and delightful to the members of the Society, but to the whole learned world.' Hooke was now forty-two, and past his prime for making new scientific discoveries, but looking forward to life as an elder statesman of the Royal. In March that year, however, the edge was taken off his happiness by his brother's suicide and Grace's 'trouble'. Preoccupied with sorting out the mess left by John Hooke, architectural work, and assisting a London publisher, Moses Pitt, with a new atlas, it was not until late July, just after his forty-third birthday, that Hooke got to grips with writing up his theory of springs for publication. The manuscript was delivered to the printer early in October, and finished books were ready by the end of November, with the title *Lectures De Potentia Restitutiva: or of Spring*. It is usually referred to by the English subtitle.

Hooke explained in the book that he had come up with the ideas it contained about eighteen years previously, 'but designing to apply it to some particular use, I omitted the publishing thereof.' The 'particular use', of course, was the spring-driven watch. The book explains how to make a simple spring balance, but goes on to point out that the same rule (or law) applies to any springy body, such as a wooden ruler placed over the edge of a table or desk and weighed down at its end: 'the force or power thereof to restore itself to its natural position is always proportionate to the Distance or space it is removed therefrom,' and 'if one power stretch or bend it one space, two will bend it two, and three will bend it three.' Twice the weight doubles the distance, three times the weight trebles the distance, and so on. This is important simply as an empirical law – early in the nineteenth century, Thomas Young refined Hooke's work by introducing a term known as 'Young's modulus', which has been described as 'the most useful of all concepts in engineering',* although few engineers appreciate Hooke's role in discovering it. Hooke himself, though, had an engineer's turn of mind, and always looked for practical applications of his work. In this case, one example he offers is that 'From this Principle it will be easie to calculate the several strengths of Bows, as of Long Bows or Cross-Bows, whether they be made of Wood, Steel, Horns, Sinews, or the like.' He also explained how a vibrating spring beats time as regularly as a swinging pendulum, the key to his work with spring-driven watches. But he went beyond the everyday practicality of his discovery. In order to explain what he had found, Hooke came up with a whole new theory of matter, in effect developing the atomic and kinetic theories decades ahead of their time.

Hooke began with the premise that the 'sensible Universe' consists of 'body and motion', and supposes 'the whole Universe and all the particles thereof to be in a continued motion'. Vibrating particles that shared the same 'harmonious' speed and amplitude of motion clung together and made up what he called 'congruous' bodies, but the size of these bodies depended on the speed with which they were vibrating, and on the pressure of their

---

* J. E. Gordon, *The New Science of Strong Materials*. Harmondsworth, London, 1976.

surroundings, harking back to his work with Boyle on air pressure. Hooke explicitly says that the particles are repeatedly colliding with one another, and that these collisions provide the outward pressure that stops them collapsing. 'All springy bodies whatsoever consist of parts thus qualified, that is, of small bodies indued with appropriate and peculiar motion.' But if the outside pressure could be increased to halve the size of the object, the frequency of the collision would be doubled, giving the object a tendency to spring back outwards. And if an object is stretched, the average distance between particles and the frequency of the collisions is correspondingly reduced, reducing the outward pressure and encouraging the object to shrink back to its original size because of the outside pressure of the air (or of the ether, a hypothetical fluid that in Hooke's day was thought to fill the Universe). He also described something akin to modern molecules: 'two or more of these particles joyned immediately together, and coalescing into one become of another nature and make a compounded particle differing in nature from each of the other particles'.

Within that framework, Hooke could also explain why bent objects tend to straighten up (because one side is stretched and the other is squeezed), and the way in which springs vibrate. Of course, the idea of atoms was not new; it went back to the time of Democritus. But the idea of atoms in random motion, colliding with one another and exerting an outward pressure that increased if they were compressed, was new, as was the suggestion in *Micrographia* that heat is simply 'a very brisk and vehement agitation of the parts of a body'. But it would be nearly two hundred years before the kinetic theory of matter became mainstream science, and by then everyone had forgotten that Hooke thought of something similar long before.

Of Spring ought to have been Hooke's swan song and scientific memorial, ensuring him of a place in the pantheon of science (if his earlier achievements had not already done that). And for a few years his status as an elder statesman of seventeenth-century science did indeed seem assured. But it would prove the calm before the storm.

As you might expect for someone at this stage of his career, Hooke's scientific work now largely consisted of refining and completing projects that he had been involved with for some time. One of the most impressive of these was a kind of automated weather station, or 'weather clock', which not only measured temperature, rainfall, air pressure and both wind speed and direction, all at the same time, but recorded the measurements by punching holes in a steadily unrolling strip of paper. The original idea went back to a suggestion by Christopher Wren in the 1650s, but it was Hooke who actually built such a weather station, drawing on his work with clocks, barometers, hygrometers and the like over the years. It was completed and demonstrated to the Fellows in May 1679. Although the device was ingenious and did work, it was rather impractical, and the punched paper recordings were difficult to interpret, so it remained just a curiosity, although a tribute to Hooke's ingenuity and practical skill.

Tantalisingly, we have only hints of what might have been a much more useful practical idea. As part of his work with the mapmaker Pitt, Hooke became interested in the process of printing, and on 13 March 1679 he noted in his diary that he had told Pitt of a 'new contrivance for printing books'. At that time, all printing presses used the flat-bed process, where each sheet of paper was laid out on a flat wooden base, with the print pressed into it by squeezing another flat piece of wood down on top of it. The process was slow, because each time the top part of the press had to be lifted, the printed page removed, and another blank page inserted before it could be printed in its turn. On 14 March, Hooke records discussing a 'contrivance for tinplates for Rolling presse', undoubtedly the idea he had mentioned to Pitt, with a friend. If this refers to a machine with a rolling cylinder pressing down on the sheet of paper being printed, Hooke was a hundred years ahead of his time: printing presses based on revolving cylinders were not developed until the 1780s. They speed things up enormously, because sheets of paper can be fed one after another into the gap between the rotating cylinder and the bed of the press, and out the other side, while the cylinder just keeps on rolling to and fro.

Although Hooke's work as City Surveyor was now largely over, he had plenty of architectural work, and was closely involved in practical aspects of Wren's work on St Paul's Cathedral at this time. In the summer, he was distracted from all these activities when Grace contracted smallpox, but unlike little Tom Giles she survived. Even for Hooke, there was too much going on for one man to handle, and something had to give. That something was the Secretaryship of the Royal, the duties of which were not suited to his abilities, and which he neglected abominably. Things got worse when Nehemiah Grew, who had been handling the correspondence of the Royal, gave up his post as joint Secretary, and virtually gave up science, in 1679, as his medical practice prospered. But the Secretaryship drew Hooke back into contact with the man who would become his nemesis: Isaac Newton.

It all started when Sir Jonas Moore, who had been the patron of the first Astronomer Royal, John Flamsteed, died in August 1679. Worried that Moore's heirs might try to claim the valuable scientific instruments at the Greenwich Observatory, the Royal quickly dispatched Hooke, along with his assistants Harry Hunt and Thomas Crawley, to recover the items, including several quadrants, which they regarded as having been loaned, not given, to the observatory. Flamsteed, left without some of his best instruments, including a quadrant made by Hooke, was livid, but could do nothing except let off steam in letters to his friends. The return of the instruments back to Gresham College, however, revived Hooke's interest in the puzzle of planetary motion, which he discussed with Wren over the autumn of 1679. On 18 October, the pair mulled over 'Elliptick motion', and on the 21st developed the theme at Bruin's coffee house. On 8 November, over coffee at Man's, Hooke told Wren about his latest ideas concerning elliptical motion 'about central attraction'. By this time, Hooke had the inverse square law clear, although he arrived at it from the point of view of a physicist, not a mathematician.

Hooke was not the only person to realise that gravity obeys an inverse square law, and he may not have been the first. But he developed a particularly neat practical explanation for why this should be the case. It is worth going into in detail because of the way it demonstrates the difference between the mind of a physicist and the mind of a mathematician.*

Hooke knew, of course, that the apparent brightness of a light, such as a candle flame, decreases as the square of its distance to the observer. A candle twice as far away appears to be only one quarter as bright. You don't need very sophisticated equipment to measure this. You could just place a single candle a certain distance away, with four identical candles twice as far away, and note that the combined light from the four distant candles was the same as the single light from the one nearby candle. This inverse square law can be explained very naturally using the wave theory of light. Imagine the light spreading out as an expanding wave in all directions from its source. This advancing wave forms a spherical shell around the source, like the skin of an expanding spherical balloon, or a soap bubble. As the sphere gets bigger, the light has to be spread in some sense more thinly, so that it can still cover the entire surface. And we know from simple geometry that the area of the surface of a sphere is proportional to the square of its radius. So when the bubble is twice as big, the area is four times as great, and at each point on its surface there will only be one quarter as much light. Hooke extended this analogy to the idea of gravity as an influence spreading out from every material object in the Universe, including the Sun, to explain why gravity also obeys an inverse square law, with the force tugging on an object being one quarter as strong at twice the distance. This is the way the physicist thinks, picturing what is going on and making analogies with similar physical systems. The mathematician plays with equations and finds the ones that match the physical reality of our world. Science progresses when the equations and the physical insight come together in one package, which is what happened with the theory of gravity. But both components are equally important, so (getting ahead of our story a little) what

---

* Hooke's version appears in the *Posthumous Works*; our adaptation contains the essence of his argument.

has come down to us as Newton's theory of gravity should really be known as the Hooke–Newton theory of gravity.

But it might not even have been known as 'Newton's' theory of gravity, had it not been for a fateful letter that Hooke wrote to Newton on 24 November 1679, when, under pressure from the Council of the Royal, he was trying to fulfil his duties as Secretary more diligently, including keeping up a correspondence with other scientists. The letter begins:

Finding by our Registers that you were pleased to correspond with Mr Oldenburg and having also had the happiness of Receiving some Letters from you my self, make me presume to trouble you with this present scribble. Dr Grews more urgent occasions having made him Decline the holding Correspondence. And the Society, hath devolved it on me. I hope therefore that you will please to continue your former favours to the Society by communicating what shall occur to you that is philosophicall, and in returne I shall be sure to acquaint you with what we shall Receive considerable from other parts or find out new here. And you may rest assured that whatever shall be soe communicated shall be noe otherwise farther imparted or disposed of then you yourself shall prescribe. I am not ignorant that both heretofore and not long since also there have been some who have indeavourd to misrepresent me to you and possibly they or others have not been wanting to doe the like to me, but Difference in opinion if such there be (especially in Philosophicall matters where Interest hath little concerne) me thinks should not be the occasion of Enmity – tis not wth me I am sure. For my own part I shall take it as a great favour if you shall please to communicate by Letter your objections against any hypothesis or opinion of mine, And particularly if you will let me know your thoughts of that compounding the celestiall motions of the planetts of a direct motion by the tangent & an attractive motion towards the centrall body . . .

Note the reasonable tone of the letter. Note also, and even more significantly, that this is the first time that Newton was made aware of the importance of the idea of an inward (centripetal) force combining with a tangential (straight line) motion to make an orbit; previously he had subscribed to the idea of an outward (centrifugal) force being constrained by something (maybe 'the ether') to stop planets flying away into space. In his reply, dated 28 November, Newton said that he had given up natural philosophy for other studies, that he had not heard of Hooke's work on springs, and that he had not come across the idea that a planetary orbit was a combination of a straight-line motion and a central attraction. 'I have had no time to entertain philosophical meditations', he wrote, and [am] 'almost wholly unacquainted with what philosophers in London or abroad have of late been imployed about.' As a result, 'believe me when I tell you that I did not, before the receipt of your letter so much as heare (that I remember) of your hypothesis.' Given Newton's reclusive nature at the time, it seems entirely possible. On the other hand, he may have been dissembling. Hooke's ideas had been published by the Royal,* and Newton had possibly seen them – he certainly seems to have admitted this in 1686, in a letter to Halley. Either way, he certainly did not realise their importance until he received the 24 November 1679 letter. As Richard Westfall, a leading Newton scholar, has put it, before that date:

> Newton's papers reveal no similar understanding of circular motion . . . Every time he had considered it, he had spoken of a tendency to recede from the centre, what Huygens called centrifugal force.'[†]

---

* In one of the Cutlerian lectures, An Attempt to Prove the Motion of the Earth, 1674, Hooke explicitly states that just as all 'Celestial Bodies' influence the Earth and one another through gravity, the Earth has a corresponding influence on their motions, and that 'all bodies whatsoever that are put into a direct and simple motion, will so continue to move forward in a streight line, till they are by some other effectual powers deflected and bent into a Motion, describing a Circle, Ellipsis, or some other more compounded Curve Line.'

† *Never at Rest.*

and:

> I do not know of any document in which Newton employed
> either the word [centripetal] or the concept before Hooke
> instructed him to do so . . . Hooke's suggestions exercised
> a profound influence on Newton's speculations [and]
> prepared his mind for the conception of universal gravita-
> tion.*

– A conception that Hooke had already had!

As these comments highlight, the correspondence initiated by
this exchange of letters triggered the development of what was
to become the most significant package of ideas in British (and
world) science – what we know now as Newtonian physics. In
1679, Newton really had given up natural philosophy for other
studies, having found the real world uncongenial after the reaction
given to his ideas about light, and was devoting his time to alchemy
and to his bizarre theological studies.† Even after Oldenburg had
persuaded him not to resign his Fellowship, he had asked the
Secretary not to forward any correspondence to him because he
intended 'to be no further solicitous about matters of Philosophy'.
He might have kept to this intention had it not been for Hooke's
letter. But in his reply, even after reiterating that he had given up
science, Newton could not resist offering his thoughts on an old
puzzle.

Ever since it had been recognised that the Earth rotates, philosophers had argued about what would happen to an object such

---

* 'Hooke and the law of universal gravitation', *British Journal for the History of Science*,
vol. 3 p. 260, 1967

† Newton was, ironically for a Fellow of Trinity College, an Arian, who rejected the idea
of Jesus as an aspect of God. If this had become common knowledge, he would have
been in deep trouble with the established Church, and forced to resign his university post.
The theological complexities need not concern us, but one result was that Newton devoted
an immense effort to studying ancient texts (or texts that were thought to be ancient) to
find evidence that he was right and the establishment was wrong. This was not just
paranoia (although Newton may well have been paranoid). His successor as Lucasian
Professor, William Whiston, was indeed dismissed after publicly proclaiming what he saw
as errors in the Anglican faith.

as a bullet, or a cannonball, dropped from a height – say, from the top of a very tall tower. Would it carry with it the forward motion associated with the rotation of the Earth, and fall at the foot of the tower? Or would it fall straight towards the centre of the Earth, get left behind by the rotation, and fall behind the tower? In fact, the question had been partially answered in 1640, when the Frenchman Pierre Gassendi arranged for balls of different weights to be dropped from the mast of a galley being rowed flat out across the Mediterranean. The balls all fell at the foot of the mast, showing that they shared the forward motion of the ship – and by implication, that they 'remembered' the motion due to the rotation of the Earth. But Newton offered another suggestion. Gassendi certainly showed that the falling balls shared the forward motion of the ship, but a mast is too short for testing the effects of the rotation of the Earth directly. The surface of the Earth at a great height, such as at the top of a mountain, is moving faster than the surface at sea level, in order to complete one revolution of a larger circle in the same time, twenty-four hours, as the lower surface. So Newton suggested that if objects could be dropped from a great enough height, and fell in a vacuum, so that there was no air resistance, they would reach the surface lower down a little further to the east, *ahead* of where they started. So far, so good. But Newton included in his letter a hastily drawn diagram to illustrate his point, with the trajectory of the falling object carrying on below the surface of the Earth and making one turn of a spiral path before reaching the centre.

After reading the scientific parts of Newton's letter out to the Royal, Hooke wrote back to Newton, trying to persuade him not to give up natural philosophy, and, perhaps in an attempt to encourage further correspondence, pointing out what he saw as a minor error in Newton's diagram. If his intention was indeed to stir Newton into scientific activity, he succeeded beyond his wildest dreams.

In this letter, written in his capacity as Secretary and duly read out to the Royal on 11 December, Hooke suggested that if it were possible for an object to fall to the centre of the Earth, it would make several ellipsoidal orbits before reaching the centre; in addition, if the falling object started out from a great height at the latitude of London it would hit the ground slightly to the southeast, not due east. This was a straightforward example of the kind of friendly scientific debate that the Royal Society encouraged, and Hooke concluded with the exhortation that Newton should 'goe on and Prosper' in natural philosophy.

But Newton was not straightforward, and reacted angrily to the letter and to the fact that Hooke, entirely in accordance with his role as Secretary, had made its content known to the Royal Society. He was still fuming about it years later, and in 1686 wrote to Halley about how it had affected him:

> Should a man who thinks himself knowing, & loves to shew it in correcting & instructing others, come to you when you are busy, & notwithstanding your excuse, press discourses upon you & through his own mistakes correct you & multiply discourses & then make use of it, to boast that he taught you all he spake and oblige you to acknowledge it & cry out injury and injustice if you do not, I beleive you would think him a man of a strange unsociable temper.*

In truth, Newton was the man 'of a strange unsociable temper'. He could not bear to be criticised or corrected, and anything that hinted at *public* criticism roused him to passion. Hooke, by contrast, was merely doing his job, and making a special effort to be seen to be doing it in the light of the justified criticism of other Fellows that he had been neglecting his duties as Secretary.

Newton's next letter, dated 13 December, was addressed not to 'my ever Honoured Friend', the form he had used before, but simply to 'Mr Robert Hooke'. In it, he went into mathematical details to show that Hooke's ellipsoidal path was also wrong. Intriguingly, a modern analysis of Newton's mathematical

---

* See Turnbull, volume 2.

methodology, carried out by Michael Nauenberg and discussed by Michael Cooper, shows that Newton used the idea of combining a tangential motion and a centripetal force in his calculation, even though he had claimed to be ignorant of Hooke's work. The fact that he went to some trouble to disguise this suggests how (un)reliable a reporter he is. In the same letter, Newton made the assumption that 'gravity be supposed uniform'. Hooke, of course, knew perfectly well that gravity obeyed an inverse square law. On 6 January, he wrote again to Newton, seemingly blissfully unaware of the hornets' nest he was poking, pointing out that the attraction of gravity falls off 'in a duplicate proportion to the Distance from the Centre', but suggesting that this inverse square law would no longer operate beneath the Earth's surface – 'not that I believe there really is such an attraction to the very centre of the Earth' – and mentioning various experiments he had carried out with pendulums and falling objects.

This gives us another insight into Newton's reliability as a reporter. In one of his letters to Halley in 1686,* he scornfully wrote that what Hooke 'told me of the duplicate proportion was erroneous, namely that it reached down from hence to the centre of the Earth.' Hooke had actually said the exact opposite; it is difficult to see this as mere forgetfulness on Newton's part, given the importance of the idea.

On 17 January 1680 Hooke wrote again in friendly fashion asking Newton to address the question of the kind of planetary orbits that would be required by an inverse square law of gravity. He seems to have been encouraging collaboration, rather than stirring up rivalry:

> It now remains to know the proprietys of a curve Line . . .
> made by a central attractive power which makes the velocitys
> of Descent from the tangent Line or equall straight motional
> all Distances in a Duplicate proportion to the Distances

---

* There was a lot of correspondence between Newton and Halley in 1686, in connection with the publication of the *Principia*.

Reciprocally taken. I doubt not that by your own excellent method you will easily find out what that Curve must be, and its proprietys, and suggest a physicall Reason of this proportion. If you have any time to consider of this matter, a word or two of your Thoughts of it will be very gratefull to the Society.

Newton did not reply, and Hooke gave up the attempt to entice him back into the scientific fold. But in another letter to Halley in 1686 Newton grudgingly acknowledged that:

his correcting my Spiral occasioned my finding the Theorem by wch I afterward examined the Ellipsis; yet I am not beholden to him for any light into yt business but only for ye diversion he gave me from other studies to think on these things & for his dogmaticalnes in writing as if he had found ye motion in ye Ellipsis, wch inclined me to try it after I saw by what method it was to be done.

'Only for ye diversion he gave me from other studies'. Overall, Hooke's time as Secretary was not a success. Administration was not his forte. But it turned Newton away from alchemy and loony theology, and back to science. Because this correspondence with Newton had such a profound influence on the development of British (and world) science, Hooke's appointment as Secretary was arguably the most significant event in his, and Newton's, life. And Newton was indeed 'beholden to him' for other ideas, not least the idea of an orbit as a combination of a straight-line motion with a centripetal attraction. It is Hooke who emerges from a reading of the correspondence as a reasonable, friendly man, eager to work with others to solve the mysteries of the Universe. As for Newton, we cannot improve on a summing up offered by Stephen Inwood:

He was neurotic, self-centred, ambitious, intolerant, over-sensitive, secretive, unforgiving and highly argumentative. It is hard to imagine Newton spending his evenings drinking coffee with a large group of congenial companions, or forming lifelong friendships with laundresses, sea captains, clerks and scientists.

With Newton having gone back into his shell at the beginning of 1680, although Hooke was now distracted, as he would have put it, from his 'other studies' and working again on science, his life settled into its usual routine. His building and mapmaking activities continued; although the details need not concern us, this meant that he was now financially independent, but his life was still centred on Gresham College and the Royal Society. From the end of 1679 to the spring of 1680, his main scientific activity involved preparing various alloys and measuring their density (or specific gravity). He found that a mixture of tin and lead is in this sense lighter than the average density of the two metals, which he explained as due to 'an aversion in the joyning of those two bodys', whereas an alloy of copper and tin was heavier than the average of the two metals because the particles (what we would call atoms) had an 'affinity' and penetrated one another. The actual measurements are less interesting to us than the emphasis Hooke put in describing them: 'Nature it selfe then is to be our Guide and we are to spend some time in her school with attention & silence before we venture to speak and teach'. In other words, experiments come first, theories afterwards.

In April 1680 Hooke made one of his rare excursions outside London to visit Lord Conway at Ragley, in Warwickshire. Hooke was designing a house for him, which is worth mentioning because although Ragley Hall was altered in the eighteenth century the basic structure remains very much as Hooke planned it.

In the summer of 1680, we get another insight into Hooke's diligence. Exasperated by the failure of the Gresham lecturers to do their duty, the authorities demanded that they should all evict their tenants, take up residence and give the lectures they were being paid for. To encourage this, they suspended all salaries. Hooke, as the only Gresham Professor who had actually been living in the college and giving his lectures (preparing them even when there was no one to hear them), appealed, and got his salary restored (with arrears paid up) because, in the words of the committee 'he only of all the lecturers hath bin constantly resident, & for ought that appears hath bin ready to read when any auditory appeared, and besides hath printed many of his lectures for the common benefit.'

The lectures presented for the common benefit in 1680 and on various occasions over the next two years concerned light; as it happens, they were not published until after Hooke's death, but in them he developed his wave theory, discussed the inverse square law for light (suggesting that this also applied to magnetism and gravity), and reiterated the importance of basing theories on experiment. It is particularly noteworthy, given the subsequent dispute with Newton, that Hooke presented the inverse square law for gravity in a public lecture in February 1680. In November that year, he spelled out his idea that the Sun exerts a gravitational force which shows up 'on the Motions of all the other primary Planets, whose Motions as I have many years showed in this Place, are all influenced and modulated by the attractive Power of this great Body'. The key words, of course, being 'attractive power', an idea that Hooke indisputably gave to Newton. In a later lecture, he said that gravity is a universal force with infinite range, acting 'on all bodies promiscuously, whether fluid or solid'. He also noted that 'comparative to the other Powers of Nature, tis weak'. After all, a child can throw a stone up into the air, against the pull of gravity.

At the end of 1680, Hooke, while still acting as Curator, was confirmed as joint Secretary of the Royal and as a Council member, while Wren was elected President, Boyle having declined the position on grounds of ill health. Around the same time, a bright comet became visible, and Hooke monitored it until it disappeared on 10 February 1681. Inevitably, this gave him the inspiration for a lecture on the history of comet studies. His workload at the Royal increased when a fellow Curator, Denis Papin, left to work in Venice, and was not replaced. But he found a new friend in the sea captain Robert Knox, who had just returned after many years in Sri Lanka (then known as Ceylon). Knox was a source of the kind of travellers' tales that Hooke loved, and Hooke helped him prepare a book, *Historical Relation of the Island Ceylon*; the two remained friends until Hooke's death two decades later.

In May 1681, Hooke lectured on the role of air in sustaining life. He had shown that a fire is only sustained by a supply of fresh air 'and without a Constant supply of that it will go out and Die'. He had also shown that fresh air, not the movement of the lungs, is essential for life: 'whether the lungs move or not move, if fresh Air be supplied, the Animal lives, if it be wanting it dies'. He was on the edge of explaining respiration as a form of combustion, a century ahead of Pierre Laplace and Antoine Lavoisier.

In the same year, Hooke studied the structure of the eye, designed an improved telescope, and built a machine with a toothed wheel that would strike a piece of metal or card and make it vibrate as the wheel turned. Changing the speed of the wheel changed the note produced by the vibration, which meant the frequency of the note could be measured. As so often with Hooke's ideas, it was soon forgotten and reinvented, this time by a Frenchman, after whom it is called 'Savart's wheel'.

With all this going on, something had to give. The diary was kept less frequently and then entries stopped altogether, and the quality of Hooke's work as Secretary also declined. There were other irritations. Flamsteed, no friend of Hooke, had become Gresham Professor of Astronomy in June 1680, providing ample scope for him and Hooke to rub each other up the wrong way during the coffee house discussions they both participated in. Flamsteed was particularly enraged when it turned out that he was wrong and Hooke was right in an argument about the behaviour of a plano-convex lens. But Hooke was not the only recipient of his wrath: Flamsteed was so disputatious that he was forced to resign in 1684, when the other professors had had enough of him. In November 1682, Hooke was not reappointed as

Secretary, and he also lost his place on the Council, although that was part of the normal rotation in which ten out of the twenty-one Council members left each year, and he would be elected to the Council again in later years. But for the time being, as far as his relationship with the Royal was concerned, at the beginning of 1683 he was once again 'only' Curator of experiments, a role he carried out alongside two other Curators, Edward Tyson and Frederick Slare. But it soon turned out that they were providing the bulk of the demonstrations, and in June Hooke's status was altered. Instead of receiving a salary, he was to be paid on results, with a quarterly assessment of what he had contributed over the previous three months.

Part of the reason Hooke had been remiss in his duties (as the Council saw it) was that his long-running dispute with Sir John Cutler had been dragging through the Courts. The tedious business resulted in Hooke receiving a payment of £200 in January 1683 and £475 in February 1684, although even then Cutler did not pay for the lectures Hooke continued to give. Freed from the hassles associated with the Secretaryship, and (at least for the time being) those associated with Cutler's reluctance to pay, Hooke once again became more active as Curator, providing experiments and demonstrations to the Fellows as before. There was nothing particularly dramatic about these, although a couple are worth mentioning. In the winter of 1683–84 he demonstrated several different kinds of accurate weighing machine, and in 1685 he carried out a detailed study of wheel design and friction, concluding that narrow wheels with a large diameter were best for the conditions of the time, rather than the wide-rimmed wheels favoured by the government and approved by regulation. He was right, but was once again ignored.

Hooke also carried out a study into the properties of ice. The second half of the seventeenth century was a period of such intense cold in Europe that it has become known as the Little Ice Age, and the winter of 1683–84 was the coldest and longest of a series of severe winters. The Thames at London froze hard enough to bear substantial weight from just before Christmas 1683 until mid-February 1684, so that it became the site of a Frost Fair, with tents and stalls laid out in streets, and both sledges and horse-drawn coaches using it as a highway. John Evelyn described it thus: '[there was] sliding with skates, a bull-baiting, horse and coach-races, puppet-plays, and interludes, cooks, tippling, and other lewd places, so that it seemed to be a bacchanalian triumph, or carnival on the water.'

Hooke left no diary for the period, so we cannot be sure that he attended the Frost Fair, but it is unlikely that he could have resisted such an attraction on his doorstep. What we do know is that he prepared a bar of ice fifteen inches long, three and a half inches thick and four inches wide, which he stood on supports placed twelve inches apart and loaded it with weights to find its breaking point. It only gave way when the load reached 350 pounds. At the meeting of the Royal on 13 February 1684, Hooke described how he had found that a piece of ice weighs only seven-eights as much as the same volume of water, so that floating ice exposes just one-eighth of its volume above the surface. This refuted the widely held notion that ice sank to the bottom when a thaw set in – the break-up of the ice on the Thames soon confirmed Hooke's assessment.

But all of Hooke's scientific activities in the 1680s would soon be overshadowed by Newton's greatest triumph. It all started innocently enough, when Hooke, Wren and Halley were discussing the inverse square law, on one of their convivial visits to a coffee shop, following a meeting of the Royal in January 1684. By then, they all knew that the inverse square law of gravity could explain the orbits of the planets. But they did not know if this was the only law that would do the job, or if anything more might be required to explain orbital dynamics. In modern parlance, was the inverse square law of gravity both necessary and sufficient to do the job? In their parlance, could Kepler's laws of planetary motion be derived from the inverse square law? Hooke said they could, but that he had not completed the calculation. Halley explained what happened next in a letter to Newton written a couple of years later:

Sir Christopher to encourage the Inquiry sd, that he would give Mr Hook or me 2 months time to bring him a convincing demonstration thereof, and besides the honour, he of us that did it, should have from him a present of a book of 40s. Mr Hook then sd that he had it, but that he would conceal it for some time that others triing and failing, might know how to value it, when he should make it publick; however I remember that Sr Christopher was little satisfied that he could do it, and tho Mr Hook then promised to show it him, I do not yet find that in any particular he has been as good as his word.

Hooke seems to have been sure that he could solve the problem, given time, so pretended that he had already solved it. In this, as we shall see, he would not be alone. But he never did solve the problem. There the matter rested, until, as we discuss in Chapter Seven, Halley raised it on a visit to Newton later in 1684. Meanwhile, Hooke worked on telescopes, barometers, his weather clock and, of course, his activities as architect and builder. He served on the Council again from December 1684 to November 1685;* at that time most of the demonstrations at the Royal were carried out by Denis Papin, who had returned from Venice, but Hooke made many contributions to the discussions. In the autumn of 1685, however, a storm in a teacup brewed up, which although initially seemed to end in Hooke's favour, would cast a long shadow.

At that time, Francis Aston and Tancred Robinson were joint Secretaries of the Royal, and responsible for the publication of the *Philosophical Transactions*. In the September–October issue of the journal they published an anonymous review of the letters of Hevelius, which raked over the old controversy with Hooke, whom the reviewer roundly criticised for:

making it his business, to carp at all [Hevelius'] Instruments, and render them suspected; to blacken and disparage to the Learned World, all his Observations

---

* In February 1685 King Charles II died and was succeeded by his brother, James II.

and repeated Hevelius' own disparaging remarks about Hooke:

> That he makes it his own business to perswade him and all
> the world, that his own way is the best, safest, and most
> exquisite, which ever can be invented by any; reproaching
> this *Author* all along for not obeying him and following his
> dictates, (as if this *Author* were one under his command;)
> Bragging only of what he can do, but doth nothing.

In the matter of open sights versus telescopic observations, of
course, Hooke's way was indeed 'the best, safest, and most exquisite'. The attack on Hooke was so unjustified that it led to a
blazing row, culminating in the resignation of Aston and Robinson,
and their replacement by John Hoskins* and Thomas Gale, both
friends of Hooke. In the aftermath of this brouhaha the Royal
also established the post of Clerk, with Halley, as we discuss in
Chapter Eight, as the first incumbent. So far, so good, as far as
Hooke was concerned. But although Newton was at that time
still in his reclusive hideaway in Cambridge and never attended
meetings of the Royal, he did read the *Philosophical Transactions*,
and took the review at face value. This would have a damaging
effect on his already shaky relationship with Hooke; the letter of
1686 mentioned above, with its reference to 'a man of a strange
unsociable temper', suggests the influence of the Hevelius
comments on Newton. It is time to cut to the chase and tell the
story of gravity from Hooke's perspective.

---

* Not the painter; his namesake.

CHAPTER SEVEN

A MISSION OF GRAVITY

The study of gravity was one of the most important scientific missions of Hooke's life – arguably, *the* most important mission. Which is why he was so upset when Isaac Newton picked up what Hooke regarded as 'his' ball and ran off with it. Hooke experimented with and studied gravity for decades, developing a sound theoretical understanding of what was going on. Then Newton came along and did a few sums that, from Hooke's point of view, confirmed what Hooke had discovered and were just the icing on the cake; and yet Newton got credit for baking the whole cake. At that time, and with some justification, the mathematical side of science was not so highly regarded as the ideas side; you need ideas, after all, before you can find the appropriate equations, as even Albert Einstein's investigation of gravity proved. So let's look at the whole story of gravity from Hooke's perspective.

Hooke's interest in gravity went back to his childhood. His early obsession with the possibility of flight – attempting to overcome gravity – was one manifestation of this, and part of his interest in developing spring-driven clocks and watches was because he knew that a pendulum clock would beat time at a different rate in places where the force of gravity was different, such as (he surmised) on top of a mountain. In the early days of the Royal Society, Hooke proposed different ways to measure gravity using falling objects, and carried out experiments with mixed results. In one series of experiments, carried out in 1663, lead weights were dropped from different heights on to one pan of a beam balance, while the other pan, which contained a heavier weight, was held in place by a light spring, to see how much what he called the 'force' of the falling object moved the balance. Descartes had argued as a principle that if an object (in physicists' language, a body) is at rest, then the impact of a smaller body will never move it, no matter how fast the smaller body moves. Hooke showed by experiment that 'the least body by an acquired celerity may be able to move the greatest'. This was the beginning of an understanding of the idea of conservation of momentum.

The next step was to measure how fast falling objects moved: in places where gravity is weaker, they would be expected to fall more slowly, so this might provide a way to measure how gravity differs from place to place. This required an accurate timekeeper, which Hooke duly built. We don't have a complete description of it, but the discussion of the experiments and their results reveal that it was a pendulum some 9¾ inches long (roughly 250 mm) which beat once every half-second. By the summer of 1664, Hooke had found that a lead ball starting from rest would drop 15½ feet in the first second of its fall. But these experiments were not followed up, partly because Hooke was kept so busy by the Royal on various projects, and partly because his investigation of gravity now took another turn.

At that time, the tallest building in London was the steeple of the old St Paul's Cathedral, and this seemed a natural place for Hooke to carry out experiments involving studies of gravity and atmospheric pressure. The cathedral had been badly damaged by fire in 1561, and in the rebuilding (completed in 1566), the steeple had been topped off with a pyramid-shaped roof, instead of a spire. A wooden platform under that roof was reachable by ladders, with a long, clear drop beneath it. Hooke went to investigate the possibilities in August 1664, and on the 25th of that month wrote to Boyle with news of his initial observations:

> One was, that a pendulum of the length of one hundred and eighty foot did perform each single vibration in no less time than six whole seconds, so that in a turn and return of the pendulum, the half second pendulum was observed to give twenty four strokes or vibrations . . . I with a plum line found the perpendicular height of [the tower] two hundred and four foot very near, which is about sixty foot higher than it was usually reported to be. In which place I shall, with some other company, this week try the velocity of the descent of the falling bodies, the Torricellian experiment, and several experiments about pendulums, and weighing.

Hooke's ambitious experimental programme was hindered rather than helped by the presence of several Fellows. But in spite of this – and in spite of the dangers and difficulties of working in a crumbling tower 200 feet above the ground balanced on an incomplete floor made of partly rotten century-old timbers, Hooke carried out many experiments over the next few weeks. His studies of gravity were inconclusive. In an earlier investigation at Westminster Abbey in November 1662, for example, he had tried weighing a piece of iron and a ball of string in a balance at ground level, then set up the balance on the roof of the abbey and used the string to lower the iron seventy-one feet while still attached to the balance, to see if its weight changed with height. He did sometimes record a small difference in weight, but nothing that could not be explained by such influences as the absorption of moisture from the air by the extended line. Similar experiments carried out a couple of years later at St Paul's also failed to find any change in the influence of gravity with height. Like his results with timing falling bodies in the tower, the results, he reported, were 'so imperfect, that I shall not, till we make them more accurate, trouble you with an account of them.'

But the opportunity to make them more accurate never arose. By October, the fading autumn light and increasingly inclement weather brought an end to the experiments. It was intended to restart the programme in the spring of 1665, but first plague and then fire meant that old St Paul's would never again be the site of scientific experimentation, and it would be a long time before the old cathedral was replaced.

As we have mentioned, by the time *Micrographia* was published in 1665 Hooke had already realised that the Moon – and by implication other celestial bodies – exerted its own gravitational influence, and during the plague year he carried out the experiments at deep wells, which we described earlier. His work so far on gravity was summarised in the report *On Gravity* that he presented to the Royal in February 1666. This concentrated on his experimental work, but was supplemented by a presentation describing his theoretical ideas about orbits, or, as he put it, 'concerning the inflection of a direct motion into a curve by a supervening attractive principle.'

> I have often wondered why the Planets should move about the Sun according to Copernicus his supposition, being not included in any solid orbs (which the Antients possibly for this reason might embrace) nor tied to it, as their Center. by any visible strings; and neither depart from it above such a degree, nor yet move in a streight line, as all bodies, that have but one singular impulse ought to doe: But all the Coelestiall bodies, being regular to solid bodies . . . must have some other cause, besides the first imprest Impulse, that must bend their motion into that Curve.

Hooke dismissed the idea that a planet trying to move in a straight line is constantly being nudged sideways by the resistance of some fluid through which it is moving, preferring the idea that the deflection is caused by:

an attractive property of the body placed in the center; whereby it continually endeavours to attract or draw it to itself. For if such a principle be supposed, all the phaenomena of the planets seem possible to be explained.

This was written in May 1666, long before Newton said anything similar; the paper is in the archive of the Royal and reprinted by Birch. It was at this time that Hooke demonstrated, using conical pendulums, how a straight-line motion could be bent into a curve by a central force.

Hooke continued to think about gravity and to make astronomical observations (among other things) even during the hectic period immediately following the Great Fire. In what became the first of his Cutlerian Lectures to be published, in 1670 Hooke described *An Attempt to Prove the Motion of the Earth*. The attempt was based on making observations at different times of year to try to detect the apparent shift in the positions of stars (the parallax) by comparing observations made on opposite sides of the Earth's orbit. The underlying principle was sound – parallax measurements are a cornerstone of modern astronomical distance measurements – but the effect is too small for Hooke to have been able to measure it with his equipment. The fact that the parallax effect is so small means that the stars are very far away, and therefore, in order to be visible at all, comparable in size to the Sun. But from our point of view, the interesting thing about this lecture is that when it was published, in 1674, Hooke included at the end a more detailed exposition of his thoughts on planetary motion. He promises that at some future time he will describe:

a System of the World differing in many particulars from any yet known, answering in all things to the common Rules of Mechanical Motions: This depends upon three Suppositions. First, That all Coelestial Bodies whatsoever, have an attraction or gravitating power towards their own Centers, whereby they attract not only their own parts, and keep them from flying from them, as we may observe the Earth to do, but that they do also attract all the other Coelestial Bodies that are within the sphere of their activity; and consequently that not only the Sun and Moon have an influence upon the body and motion of the Earth, and the Earth upon them, but that Mercury also Venus, Mars, Jupiter and Saturn* by their attractive powers, have a considerable influence upon its motion as in the same manner the corresponding attractive power of the Earth hath a considerable influence upon every one of their motions also. The second supposition is this, That all bodies whatsoever that are put in a direct and simple motion, will so continue to move forward in a streight line, till they are by some other effectual powers deflected and bent into a Motion, describing a Circle, Ellipsis, or some other more compounded Curve Line. The third supposition is, That these attractive powers are so much the more powerful in operating, by how much nearer the body wrought upon is to their own Centers. Now what these several degrees are I have not yet experimentally verified; but it is a notion, which if fully prosecuted as it ought to be, will mightily assist the Astronomer to reduce all the Coelestial Motions to a certain rule, which I doubt will never be done true without it.

Michael Cooper has commented that this 'resembles very closely Newton's world-view as it eventually appeared seventeen years later'. It would be more accurate to say that the world-view presented in the *Principia* resembles very closely Hooke's world-view as it appeared seventeen years earlier![†] There is everything here except the inverse square law, and Hooke soon had that, as well. As we have seen, by about 1676 Hooke, Halley and Wren were aware that an inverse square law of gravity would do the trick (at least for circular orbits; the more general application was

---

* Hooke used the astrological symbols for the planets, rather than spelling out their names.

† And remember that Newton did eventually acknowledge to Halley that he had read *An Attempt to Prove the Motion of the Earth*.

tricky to prove). In 1677, Hooke observed a comet and at the behest of the Royal published a paper the following year,* drawing on this and his earlier observations for comets. Quoting from his lecture notes for 1665, Hooke posed the key questions concerning the orbit of a comet:

> What kind of motion it was carried with? Whether in a straight or bended line? And if bended, whether in a circular or other curve, as elliptical or other compounded line, whether the convex or concave side of the curve were turned towards the Earth? Whether in any of those lines it moved equal or unequal spaces in equal times? [and] Whether it ever appears again, being moved in a circle; or be carryed clear away, and never appear again, being moved in a straight or paraboloeical line?

He reached the conclusion that cometary nuclei contain solid matter that possesses its own 'gravitating principle' so that it is attracted towards the Sun and deflected (as he put it, 'incurvated') from a straight line into a curved path around the Sun, although in this case 'it were not wholly stayed and circumflected into a circle'.

The scene was set for the correspondence that Hooke initiated at the end of 1679 in his capacity as Secretary, culminating with his statement to Newton in the letter of 6 January 1680 that 'the Attraction always is in a duplicate proportion to the Distance from the Center Reciprocall.'

So by 1680 Hooke had not only worked out, but had presented to Newton, a complete world-view incorporating the idea of universal gravitation, the first law of motion (that every body stays at rest or proceeds in a straight line at constant speed unless deflected by a force) and the centripetal inverse square law. He pointed out that the force of gravity should be calculated in accordance with the inverse square law as if it acted from the

---

* As *Cometa*.

centre of a body such as the Earth or the Sun, but that the inverse square law did not work below the surface of such a body. He knew that the Universe was governed by physical laws, the same laws that applied here on Earth, and not by mystic powers. Before he received this package of ideas, Newton's world-view was very much what you would expect from a mystic alchemist and crackpot theologian. He thought that the planets were kept apart by 'unsociableness' and referred to vortices in the ether. On 7 December 1675, a year after the publication of *An Attempt to Prove the Motion of the Earth*, he wrote to Oldenburg spelling this out:

> So may the gravitating attraction of the Earth be caused by the continual condensation of some other such like ethereal Spirit, not of the maine body of flegmatic aether but of something very thinly and subtily diffused through it, perhaps of an unctuous or Gummy, tenacious & Springy nature, and bearing much the same relation to aether, we the vital aereall Spirit requisite for the conservation of flame & vitall motions (I mean not ye imaginary volatile saltpeter), does to Air. For if such an aethereall Spirit may be condensed in fermenting or burning bodies, or otherwise inspissated in ye pores of ye earth to a tender matter wch may be as it were ye succus nutritious of ye earth or primary substance out of wch things generable grow (or otherwise coagulated, in the pores of the earth and water, into some kind of humid active matter for the continuall use of nature. adhereing to the sides of those pores after the manner that vapours condense on the sides of a Vessell subtily set); the vast body of the Earth, wch may be every where to the very centre in perpetuall working, may continually condense so much of this Spirit as to cause it from above to descend with great celerity for a supply.

Phew! After rambling on a bit more, Newton gets to the question of how the Sun and the planets stay in their separate places:

So some fluids (as Oyle and water) though their pores are in freedome enough to mix with one another. Yet by some secret principle of unsociablenes they keep asunder, & some that are Sociable may become unsociable by adding a third thing to one of them, as water to Spirit of Wine by dissolving Salt of Tartar in it. The like unsociablenes may be in aethereall Natures, as perhaps between the aethers in the vortices of the Sun and Planets.

Comparing those almost contemporaneous accounts, the modern reader is left in no doubt who was the forward-looking scientist with great insight, and who was the backward-looking mystic with a head filled with magical mumbo jumbo.* But if this was the confused state of Newton's thinking about gravity and planetary orbits in the second half of the 1670s, what does it tell us about the famous story of the falling apple, which according to Newton he saw in 1665 or 1666, and made him realise that the same force that pulls the apple to the ground holds the Moon in its orbit, and gave him the inspiration for his theory of gravity, including the inverse square law? Simple. He made it up, in order to ensure that posterity would not realise how much he got from Hooke.

The story as it has come down to us dates from an anecdote he told in 1726, the year before he died. The story was told to William Stukeley, a much younger man who came from the same part of the country and collected anecdotes about Newton in a hero-worshipping and uncritical way. Although this particular story dates from 1726, it was not written down until 1752. The manuscript is in the Royal Society archive and can be accessed online, but the story appears in Stukeley's *Memoirs of Sir Isaac Newton's Life*, edited by A. H. White and published in 1936. According to Stukeley, on 15 April 1726:

---

* For comparison, remember that in his diary entry for 25 November 1678, Hooke referred to astrology as 'vaine'.

After dinner, the weather being warm, we went into the garden & drank thea under the shade of some apple tree; only he & myself.

Amid other discourse, he told me, he was just in the same situation, as when formerly the notion of gravitation came into his mind. Why shd that apple always descend perpendicularly to the ground, thought he to himself; occasion'd by the fall of an apple, as he sat in contemplative mood.

Why shd it not go sideways, or upwards? But constantly to the Earth's centre? Assuredly the reason is, that the Earth draws it. There must be a drawing power in matter. And the sum of the drawing power in the matter of the Earth must be in the Earth's centre, not in any side of the Earth.

Therefore does this apple fall perpendicularly or towards the centre? If matter thus draws matter; it must be proportion of its quantity. Therefore the apple draws the Earth, as well as the Earth draws the apple. There is a power, like that we here call gravity, which extends its self thro' the universe.

This was clearly a story that Newton had rehearsed and polished. He also told it to John Conduitt, the husband of Newton's niece, who recorded:*

In the year 1666 he retired again from Cambridge to his mother in Lincolnshire. Whilst he was pensively meandering in a garden it came into his thought that the power of gravity (which brought an apple from a tree to the ground) was not limited to a certain distance from Earth, but that this power must extend much further than was usually thought.

Why not as high as the Moon said he to himself & if so, that must influence her motion & perhaps retain her orbit, whereupon he fell a calculating what would be the effect of that supposition.

---

* http://www.newtonproject.sussex.ac.uk/view/texts/normalized/THEM00167

Apart from the evidence of Newton's own letter to Oldenburg in 1675, which gives the lie to the account, there is an even more striking example of Newton's attempt to rewrite history. In his correspondence with Halley, he refers to a letter sent to Oldenburg in June 1673, to be copied and sent on to Huygens, in which he spells out the idea of a centripetal gravitational attraction. Sure enough, the version of the letter in the archive of the Royal supports this claim. But, alas for Newton's reputation, the letter in the Huygens archive does not contain the relevant material. The version held by the Royal has been faked.*

So the fabled story of the fall of an apple in the plague year of 1665 giving Newton the inspiration for 'his' theory of gravity is just that – a fable. Until he received Hooke's letter, Newton had not realised that gravity is a universal force that affects all objects, but had thought of orbital motion in terms of vortices or whirlpools in some mysterious fluid, carrying planets round the Sun (and the Moon around the Earth) like chips of wood floating in vortices in a river. This is not a system that readily explains the fall of an apple from a tree. By the time he came up with the apple story, Newton was an old man and the plague year was a distant memory (and, of course, Hooke was dead). There are two possible reasons why he told the story. The charitable version is that over the years he had first hit on the falling apple and the falling Moon analogy as a neat demonstration of universal gravity at work, then convinced himself that this must have been how he got the idea in the first place. Old men reminiscing about their youth do tend to make mistakes like this. But the evidence suggests that it would be a mistake to put a charitable interpretation on anything Newton said, especially where it concerned his priority vis-à-vis Hooke. The other possibility is that Newton deliberately made the story up, and told it to make sure that future generations would never guess that Hooke deserved the credit for the concept of universal gravity. This seems to us much more likely, given the

* Louise Patterson, 'Hooke's Gravitation Theory and Its Influence on Newton, II', *Isis*, vol 41, pp 304–305, 1950.

clear evidence that Newton told lies about his priority claims on more than one occasion. Either way, we have only Newton's word, the word of a known liar, that the falling apple story is true; which reminds us that there were good reasons why the Royal Society chose as its motto *nullius in verba*.

Newton does not seem to have latched on immediately to Hooke's ideas concerning a universal centripetal attractive force of gravity in 1680, but still clung to his idea of vortices. But a comet that appeared in 1682 did not fit that world-view, and is probably the reason why he started to take Hooke's suggestion seriously. This is the object now known as Halley's comet, which Hooke, like other astronomers, studied during August and September that year. As well as its later significance to our story (see Chapter Eleven), it has one important property: it goes round the Sun in the opposite way to the planets – a so-called retrograde orbit. This simply does not fit the image of all the bodies in the Solar System being swept around the Sun in a swirling fluid – the comet would have to be moving 'upstream', against the flow of the 'aethereall Spirit'. So Newton was primed not only by the correspondence with Hooke but also by the fact of retrograde cometary orbits when Halley happened to visit Cambridge in 1684 and took the opportunity to ask Newton for his thoughts about orbits and the inverse square law.

Newton's personal circumstances had changed around this time. From his early days in Cambridge, Newton had shared a set of rooms in Trinity with another scholar, John Wickins, who among other things had acted as his scribe. Wickins resigned his Fellowship in 1683, became Rector of Stoke Edith in Hertfordshire, and married. Newton seems to have been somewhat disgruntled by the break-up of their friendship, but in 1685 Wickins was replaced by a young man just up from Grantham, Isaac's namesake Humphrey Newton (neither of them ever claimed that they were related). Humphrey only stayed with Isaac for five years, but it is worth mentioning because, among other things, he wrote out the fair copy of Newton's masterwork, the *Principia*, from which the printer worked.

A few months after Wren had challenged Hooke and Halley to find a proof of the inverse square law, with the inducement of a book prize, Halley had to travel to Peterborough, in the summer of 1684, on the family business we discuss in Chapter Eight. He took the opportunity to take a slight detour to Cambridge to meet up with Newton, where he raised the subject that had been nagging at Hooke, Wren and himself since January. What is an undoubtedly embellished account of what happened next (but the only account we have) comes from Abraham de Moivre, a French Huguenot refugee who was an acquaintance of Newton and reported what Newton told him of the events:

> In 1684 Dr Halley came to visit him in Cambridge, after they had been some time together the Dr asked him what he thought the Curve would be that would be described by the Planets supposing the force of attraction towards the Sun to be reciprocal to the square of their distance from it. Sr Isaac replied immediately it would be an Ellipsis, the Dr struck with joy & amazement asked him how he knew it, why saith he, I have calculated it, whereupon Dr Halley asked him for the calculation without any further delay, Sr Isaac looked among his papers but could not find it, but he promised to renew it, & then send it to him.*

Just like Hooke in January, Newton was playing for time by pretending he had already made the calculation, and planning to do it once Halley was out of the way, so that he could then claim to have 'found' the missing piece of paper. But unlike Hooke, Newton did have the mathematical skills to carry out the proof, as well as both the opportunity and inclination to work single-mindedly (even obsessively) on one project, without being distracted by other activities.

The initial result of this single-minded devotion to the problem

---

* Writing long after the event, de Moivre gives Newton and Halley the titles ('Sir' and 'Dr') that they did not hold in 1684.

of planetary orbits was a short paper titled *De Motu Corporum in Gyrum* (*On the Motion of Bodies in Orbit*), which Newton sent to Halley in November 1684. This was only an outline of Newton's ideas, but enough to show Halley that Newton should be encouraged to flesh it out into a complete package. From our point of view, the most significant thing about *De Motu*, as it is usually known, is that in it Newton introduces the term centripetal force, having at last taken on board Hooke's realisation that orbital motion is a result of the combination of the tendency of a body to move in a straight line and an inward (centripetal) pull of gravity. In *De Motu*, Newton showed that under the influence of a centripetal inverse square law planets would indeed follow elliptical orbits obeying Kepler's First Law, that an imaginary line joining the planet to the Sun sweeps out equal areas in equal intervals of time.

Halley returned to Cambridge to discuss the paper with Newton, and then, on 10 December, announced its existence, but probably not its detailed contents, to the Royal. He referred to it as a 'curious treatise', by which he meant that it invited curiosity, not that it was strange. Hooke was, therefore, at least aware of *De Motu*, and Halley may have given him a chance to read it. But he was not prompted by this to return to his own studies of orbital motion, and seems to have accepted that he had taken this as far as he could.* But it must have been clear to him how much his ideas had shaped Newton's thinking, and he no doubt expected credit for them in due course.

Halley now became in effect the midwife for Newton's masterpiece. Enthusiastically backed up by the Royal, he encouraged Newton to concentrate on writing what became *Philosophiae Naturalis Principia Mathematica* (*The Mathematical Principles of Natural Philosophy*) – the *Principia*. Alchemy and theology were pushed to one side, while Halley encouraged Newton when he tired of the project and soothed his ruffled feathers when he got

---

* He did draft a paper on 'The Laws of Circular Motion' in the late summer of 1685, but it did not go as far as *De Motu*, and he never published it.

angry, particularly about Hooke's claims. More than a year after Halley had started cajoling Newton to develop the ideas hinted at in *De Motu*, the first fruit appeared. On 21 April 1686, Halley, by then Clerk of the Royal, told the Society that Newton's book was nearly ready for publication, and the following week the manuscript of Book One of the *Principia* arrived in London and was presented to them. It claimed to explain 'all the phaenomena of the celestial motions by the only supposition of a gravitation towards the center of the sun decreasing as the squares of the distances therefrom reciprocating'. It all sounded very familiar to Hooke.

We know of the immediate fallout from the meeting of 28 April 1686 from a letter that Halley later wrote to Newton. The meeting was chaired by Sir John Hoskins, who said that Newton's book 'was so much more to be prized, for that it was both Invented and perfected at the same time.' We can imagine Hooke sitting there fuming, thinking that it may have been perfected by Newton, but it had been invented, years before, by himself. Exactly what he said to Hoskins we do not know, but Halley tells Newton (and us) that the two of them 'who till then were the most insep-arable cronies, have since scarce seen one another, and are utterly fallen out.' At the usual coffee house gathering after the meeting, Hooke tried to press the point that he had long had both the inverse square law and the idea of a centripetal attraction, but to no avail.

Halley was a good friend of Hooke, but his immediate objec-tive was to see the *Principia* through to publication.* When Halley wrote to Newton on 22 May 1686 to inform him that the Royal intended to pay for the cost of printing the book, he made an attempt to pre-empt trouble by giving an undramatic account of these events to Newton before some wilder story might reach him, with a carefully phrased attempt to maintain equilibrium between Newton and Hooke:

---

* Indeed, he was so committed to the project that when the Royal ran short of funds Halley in the end paid for its publication himself; happily, he probably made a small profit when the books were sold.

There is one thing more that I ought to inform you of, viz, that Mr Hook has some pretensions upon the invention of ye rule of the decrease of Gravity, being reciprocally as the squares of the distances from the Center. He sais you had the notion from him, though he owns the Demonstration of the Curves generated therby to be wholly your own; how much of this is so, you know best, as likewise what you have to do in this matter, only Mr Hook seems to expect you should make some mention of him, in the preface, which, it is possible, you may see reason to praefix. I must beg your pardon that it is I, that send you this account, but I thought it my duty to let you know it, that you so may act accordingly; being in myself fully satisfied, that nothing but the greatest Candour imaginable, is to be expected from a person, who of all men has the least need to borrow reputation.

At first, Newton responded as Halley had hoped. He agreed that Hooke had made a contribution to the understanding of planetary motion, and that Newton would give him credit where due. But in a blatant lie he said that nothing in the correspondence of 1679–1680 had contributed to Newton's thinking, and had the chutzpah to say that if his recollection was at fault 'I desire Mr Hooke to help my memory'. Perhaps Newton's memory really was that bad. As he brooded in Cambridge, he seems to have grown increasingly angry about Hooke's claims, an anger fuelled by a more colourful account of Hooke's clash with Hoskins, supplied by Edward Paget, who was both a Fellow of the Royal and a Fellow of Trinity, Newton's college. On 20 June 1686 Newton wrote a much more intemperate letter to Halley, incorrectly saying that Hooke had claimed the inverse square law operates all the way to the centre of the Earth (which Hooke had explicitly said was not the case) and making unjustified claims that Hooke had actually stolen his ideas, such as they were, from other people. But he did make one valid point, albeit in a rather over-the-top way. It was all very well having bright ideas, but somebody had to do the work of proving them.

Now is this not very fine? Mathematicians that find out, settle & do all the business must content themselves with being nothing but dry calculators & drudges & another that does nothing but pretend & grasp at all things must carry away all the invention as well as of those that were to follow him as of those that went before. Much after this same manner were his letters writ to me . . . And upon this information I must now acknowledge in print I had all from him & so did nothing my self but drudge in calculating demonstrating & writing upon ye inventions of this great man.

Of course, Hooke did not wish to 'carry away all the invention', but simply to have credit for his work, and if anyone, it was Newton who was carrying away the invention of 'those that went before'.

Halley did not read this letter out to the Royal, but filed it away quietly. Newton had also threatened to withhold Book Three of the *Principia*, which was already in draft form, from publication. So Halley went into soothing overdrive. He now (29 June) told Newton about the coffee house discussion of January 1684, the bust-up with Hoskins, and the circumstances surrounding Hooke's claim, which, he said, had been 'represented in worse colours than it ought'. Halley succeeded in soothing Newton to the extent that he admitted on 14 July 1686 that 'he [Hooke] was in some respects misrepresented to me', withdrew his threat to withhold Book Three, and conceded the minor point that the correspondence with Hooke had led him to seek a mathematical basis for an understanding of the motion of the planets. But the overall outcome of this brouhaha was that Hooke's work was not acknowledged as fully in the *Principia* as it would have been if he had bitten his lip and kept quiet.

In the draft of Book Three, which Halley had not yet seen, Hooke was mentioned as one of the discoverers of the importance of centripetal force, along with 'others of our nation'; this was already minimising his contribution, since there were no such 'others', but in one of his fits of rage Newton crossed the reference out. He then went through the manuscript savagely erasing almost every reference to Hooke (he may have overlooked some of the ones that remain), and where he could not ignore Hooke's contribution to observations of comets he reduced the reference from *Clarissimus* ('the very distinguished') *Hookius* to plain *Hookius*. But he did acknowledge that 'our countrymen Wren, Halley and Hooke' had been aware of the inverse square law.

Newton also made some other changes to Book Three. He had originally drafted it in what he called 'a popular method' to make the ideas relatively easily accessible. But he changed it into a mathematical treatise of deliberately great complexity. He later told William Derham, who was a Fellow of the Royal and an acquaintance of Newton (and eventually edited some of Hooke's papers for publication), that he had done this 'to avoid being baited by little smatterers in mathematics'. We can guess which 'little smatterer' he had in mind. In the final version of Book Three Newton also added Proposition XIX, suggesting that the Earth bulges at the equator because of its rotation; Hooke, who had made this suggestion some time earlier (as Newton was well aware), received no acknowledgement. But, as 'Espinasse has pointed out, after delivering the book to the Royal Newton took no interest in its publication, and as he had written to Oldenburg in 1676 did now 'bid adieu' to science.

Publication of the *Principia* was completed in July 1687, establishing Newton in the eyes of the world as the greatest scientific giant. Michael Cooper has suggested that 'without Hooke's hints and goading and Halley's cash and moderating influence on both men, Newton might not have bothered with it.' But he did bother, and Hooke would soon be relegated to a footnote of scientific history. The dispute rumbled on, at least from Hooke's side, for the rest of the century. But in 1687 he was also struck by a more devastating personal blow when Grace died, at the age of only twenty-seven (Hooke was then fifty-two). Richard Waller tells us that 'the concern for whose Death he hardly ever wore off, being observ'd from that time to grow less active, more Melancholly and Cynical'.* More of Hooke's declining years shortly, but first we want to spell out the key insights in the *Principia*, which paved the way for the development of modern physics, and highlight just who deserves credit for what.

First, we have the idea of a universal law of gravitation, the same in heaven as on Earth, showing that the same laws apply throughout the Universe. Definitely Hooke. Then, the fact that gravity is a centripetal attraction. Hooke again.† The proof that an inverse square law is necessary and sufficient to explain planetary motion. Newton, but prompted by Hooke. Newton also spelled out in his book three laws of motion, which formed the foundations of physics. First, every body continues at rest or moving in a straight line unless acted upon by a force. Hooke, of course. Second, the acceleration of an object is proportional to the force acting on it. Newton. Third, when a force is applied to a solid object it responds with an equal and opposite force – action and reaction. Newton. And a more subtle but crucial notion: gravity operates at a distance, without the need for any intervening fluid such as the ether. Hooke.

By our count, that is Hooke four, Newton three, but allowing the importance of the mathematics we might as well call it fifty–fifty. That is, Hooke made a fifty–fifty contribution to Newton's singular achievement, as well as all the other things he did for the development of science, while Newton made a fifty–fifty contribution to his own singular achievement, and nothing else for the development of science, except to make pretty patterns with prisms.

We exaggerate slightly, but nowhere near as much as the achievements of Newton have been exaggerated, relative to the

---

* This was something of an exaggeration, as we shall see in Chapter Nine, but contained a grain of truth.

† Richard Westfall has commented 'without the concept of centripetal force, the theory of universal gravitation was inconceivable and Hooke's contribution to gravitation was not then insignificant' ('Hooke and the Law of Universal Gravitation', *British Journal for the History of Science*, volume 3 pp 245–261, 1967). This from a Newton biographer with no axe to grind for Hooke.

achievements of Hooke, for more than three centuries.

Remember Hooke's practical explanation for why the inverse square law should apply to gravity, based on the idea of an influence spreading out over the surface of an expanding sphere. As we explained, it also demonstrates the difference between the mind of a physicist and the mind of a mathematician. The physicist pictures what is going on and makes analogies with similar physical systems. The mathematician plays with equations and finds the ones that match the physical reality of our world. Science progresses when the equations and the physical insight come together in one package, but in the mid-seventeenth century there was no sense that the mathematical approach was more important, let alone pre-eminent. Indeed, ideas were the key component. Hooke was a great scientist, who came up with the first scientific world-view; Newton was a great mathematician, who put Hooke's world-view on a secure mathematical foundation, and then claimed credit for the whole thing himself.

Although this is not a book about Newton (he has had plenty of books written about him), it would be remiss not to mention what happened to him after the publication of his masterwork, not least because his life took such an unexpected turn.

Postscript: Newton after the *Principia*

Newton did little of scientific significance after the publication of the *Principia*, although when Hooke was safely out of the way he did publish his *Opticks*, based on work he had done decades before and also including, without credit, Hooke's ideas. A colleague of ours used to joke about a reference he had received in support (?) of an applicant for an academic post. It said 'he is a man of singular talent'. Our colleague claimed that this meant 'he only had one good idea in his life'. It might be slightly unkind to apply this epithet to Newton, but it isn't far from the truth, especially compared with the breadth and depth of Hooke's contributions. So what did Newton get up to in the forty years from 1687 to 1727?

The dramatic change in Newton's life began before the publication of the *Principia*, and was a result of the death of Charles II in 1685 and the accession of his brother, James II. Charles had been happy to be at least nominally Protestant; James, however, was not only openly Catholic, but was foolish enough to attempt to impose his Catholicism on British institutions, including the University of Cambridge. On 9 February 1687, he ordered the University to admit a Benedictine monk, Alban Francis, to the degree of Master of Arts. This kind of thing was not uncommon – rather like the modern practice of honorary degrees. But it was clear that Francis, if he was admitted, would exercise his right as an MA to participate in the affairs of the University, and that James intended to make further appointments of this kind, packing the administration of the University with Catholics.

The University resisted, and one of the leaders of the resistance was Newton, who suddenly came out of his shell and became actively involved in politics. The reason was simple. He was a closet Arian who did not accept key doctrines of the Church. By keeping quiet and, among other things, not taking Holy Orders (very unusual for a University Fellow), he had been able to get by as nobody had asked him any awkward questions. But if the University were to be taken over by Catholics, those awkward questions would surely be asked, and at the very least he would lose his position, as would indeed later happen to William Whiston. Attack being the best form of defence, Newton now devoted himself to the attack, speaking out against James's interference in University affairs. This took some courage: Newton was one of nine Cambridge Fellows summoned before the notorious Judge Jefferies, then Lord Chancellor, to explain himself. But James's position was increasingly shaky, and he was in no position to enforce his wishes. There was considerable Catholic sympathy in the country at large, but there was also an overwhelming wish not to get involved in any re-run of the turmoil of the middle part of the seventeenth century.

Something had to give, and the something turned out to be James. At the end of 1688, England was invaded by an army under the Dutch Prince William of Orange, who had been encouraged by a large section of the English Parliament to take the throne. William was the son of Charles II's sister, and his wife Mary was the daughter of James II, which lent some semblance of legitimacy to the invasion, which was afterwards referred to by seventeenth-century spin doctors as 'the Glorious Revolution'. But don't be fooled by the spin: this, not the events of 1066, was the last successful invasion of British soil. James was allowed to leave quietly for exile on the continent.

In order to legitimise the takeover, a Convention Parliament was called in 1689 to establish William and Mary as joint monarchs, and the Church of England as, well, as the Church of England. The University of Cambridge was entitled to send three of its Fellows as Members of Parliament (nothing so sordid as a public election was involved), and Newton, wreathed in the glory of his public stand against James, was one of them. He said nothing and voted the party line during the year and a month that the Parliament sat, but suffered one disappointment. When the Parliament passed an Act to give more tolerance to religious dissenters, it specifically excluded 'any person that shall deny in his Preaching or Writeing the Doctrine of the Blessed Trinity.' But the disappointment must have been tempered by a development in Newton's personal life.

Christiaan Huygens had a brother, Constantijn, who was a secretary to King William. Naturally, Christiaan visited his brother in London and attended meetings of the Royal, where he met Newton on 12 June 1689. On 9 July Newton visited Christiaan at Hampton Court. On both occasions a young Swiss mathematician, Nicholas Fatio de Duillier (usually referred to as Fatio), was also present, as a member of Huygens' party. Fatio seems to have hero-worshipped Newton, in the manner of a scientific groupie, and Newton seems to have become infatuated with Fatio. The intense friendship lasted for about four years. Fatio spent a month with Newton in London in 1690, then just over a year based in the Netherlands, where he acted as a conduit for the exchange of ideas between Newton and Huygens. There were other visits to England, and plans for Fatio to live in Cambridge while working with Newton on a new edition of the *Principia*. But the close relationship ended in 1693. After Fatio suffered an unspecified illness at the end of 1692, Newton urged him repeatedly to come to Cambridge to recuperate, but Fatio declined and the correspondence petered out, although Fatio remained an extravagantly enthusiastic supporter of Newton, particularly in the dispute with Leibniz about who had invented calculus.

In the early 1690s, Newton found life in Cambridge frustrating after his experiences on the broader London stage, and began looking for other employment while devoting himself once again to alchemy. Overwork, frustration at the lack of new opportunities, the break-up of his friendship with Fatio, and the trial of keeping quiet about his sexuality and religious thinking all took their toll, and he suffered a major nervous breakdown in 1693. He gradually recovered, with the support of his colleagues, and in 1696 was offered the post of Warden of the Royal Mint, which he seized on with delight. The appointment was officially dated from 19 March, and by the end of April Newton had shaken the dust of Cambridge from his shoes and was installed in London, where he would remain (except for a few brief visits to Cambridge) for the rest of his life.

The appointment was clearly intended as a sinecure, a reward for past services; the letter of appointment said that it did not 'require more attendance than you may spare'. The Warden was Number Two at the Mint, which was officially run by the Master. But Newton was incapable of taking things easy, and the Master, Thomas Neale, was happy to let him do all the work if he was so inclined. And there was plenty of work to do. The Mint was about to carry out a complete recoinage, a task that Newton oversaw with his usual single-minded obsessiveness. We also get another insight into his character. One of his tasks was to prosecute counterfeiters, a job he relished so much that he became a magistrate, so that he could not only catch forgers but convict them, usually to be hung. When Neale died in 1699, Newton took over as Master of the Mint, a post he held until he died, although in later years his work was done by a deputy. In 1701, Newton resigned all his posts in Cambridge, having made £3,500 from his post at the Mint that year.

History repeated itself in the early 1700s. Newton became an MP again in 1701, and when William died in 1702 (Mary had predeceased him), Queen Anne, Mary's sister, succeeded to the throne. During an election campaign in 1705, Anne gave knighthoods to some of her favourites, including Newton, to encourage support for the faction she favoured. The faction Anne favoured did poorly, however, and he was not returned to Parliament that year, or ever again. But the tale is worth telling because many people think Newton was the first person to be knighted for his scientific work. He was not: he was knighted as a party political ploy.

Although Newton would never again serve as an MP, he now had something else to occupy him, apart from his duties at the Mint. In his early sixties in 1703, and with Hooke dead, he was elected as President of the Royal Society, which he ruled with a rod of iron for the next two decades. He also took the opportunity to publish his *Opticks* in 1704. And somehow, early in Newton's Presidency when the Royal moved out of Gresham College and into new quarters at Crane Court, the only known portrait of Robert Hooke disappeared. Newton died on 20 March 1727, at the age of eighty-two. Plenty of portraits of him survive. It is worth noting, though, that as late as 1717, in the second English edition of the *Principia*, Newton, the mystic, was still trying to 'explain' gravity in terms of the influence of a medium, the aether, with non-uniform density, through which the planets ('gross bodies') moved. His 'Query 21' is as confused as any of his writing before the correspondence with Hooke about planetary motion:

> And so if any one should suppose that aether (like our air) may contain particles which endeavour to recede from one another (for I do not know what this aether is), and that its particles are exceedingly smaller than those of air, or even than those of light, the exceeding smallness of its particles may contribute to the greatness of the force by which those particles may recede from one another, and thereby make that medium exceedingly more rare and elastick than air, and by consequence exceedingly less able to resist the motions of projectiles, and exceedingly more able to press upon gross bodies, by endeavoring to expand itself.

It seems he never did grasp the significance of Hooke's concept of a centripetal force which deflects the straight line path of a planet (or, indeed, a comet) into a curve. The irony will become clear in Chapter Eleven, but, as we have hinted, we might well have never had the *Principia* at all, had it not been for a series of unfortunate events.

CHAPTER EIGHT

HALLEY, NEWTON
AND THE COMET

Isaac Newton may never have produced his masterwork if Edmond Halley's father had not been murdered. Since that murder has been linked with the sordid aftermath of a plot against the King and his brother, it seems that we may not have had the *Principia*, with all that that implies for the history of science, had it not been for the Catholic leanings of Charles II, and the open Catholicism of James, Duke of York, the heir apparent.

The anti-Catholic plotters intended to murder both Charles and James at Rye House, Hoddesdon, in Hertfordshire, on their way back to London from the races at Newmarket in April 1683; so the misadventure has become known as the Rye House Plot. The plot was betrayed, and several conspirators, including the Earl Russell, Algernon Sidney, and the Earl of Essex, were arrested and imprisoned in the Tower of London awaiting trial for high treason. Russell and Sidney were duly tried, convicted and beheaded. But Essex died in the Tower, without coming to trial, in mysterious circumstances, which is where Edmond Halley senior comes into the story.

As a prominent citizen, the elder Edmond Halley was a Yeoman Warder of the Tower. This was a largely honorary position, with some ceremonial duties, and with the advantage that it excused holders of the post from certain other tedious tasks. Alan Cook, who provides the most complete and up-to-date account of the circumstances surrounding the elder Halley's death, quotes a pamphlet from 1690 which says 'This Mr Hawley was very rich, and a warder only to exempt him from parish services but he never waited [at the Tower], unless it were on very solemn occasions'. On 13 July 1683, both the King and the Duke of York visited Essex in the Tower, and Halley senior was among the party, this no doubt being a solemn occasion, if not part of the usual ceremonial duties. Shortly afterwards, Essex died, apparently having cut his own throat. The official story was that he asked for a small knife to trim his fingernails, but that the guard did not have such a knife so lent him a razor. Essex then went into an annexe off his main apartment, where he was later found with his throat slit and covered in blood. Although there were inevitably rumours that he had been murdered, suicide does seem the most likely bet. But the guard seems to have been remarkably careless, and the most plausible explanation of the course of events is that the King, or more probably James, colluded with Essex to allow him to avoid the stigma of conviction and beheading. The Earl's father had died fighting for Charles I, which might have encouraged leniency, so Essex might have been offered an easy way out, according to the code of the time. Had he been convicted of treason, apart from the stigma there was the practical matter that the family estates would have been forfeited to the Crown. There is also a possibility that Essex was indeed murdered. We shall never know the details, but Edmond Halley senior certainly seems to have known something, which probably cost him his own life.

On 5 March 1684, the elder Halley left home in the morning, telling his wife that he would be back in the evening. He had been having a problem with shoes that pinched his feet, and removed the lining before he left the house. He was never seen alive again. More than a month later, on 12 April, his body was found washed up near Rochester, on the Medway. It was badly disfigured, although the exact details of the injuries were never published, and identified by the victim's unlined shoes. But the inquest concluded that he had not been in the river long, or the body would have been more 'corrupted'. The verdict was murder by some unknown person or persons, and it seems likely that the corpse was thrown overboard from an outward-bound ship. Speculation at the time, and for years afterwards, was that the elder Halley knew too much about the fate of Essex, had been brooding for months and was now threatening to talk, so he had been silenced. Conspiracy theorists had a field day, with pamphlets being published about the case, but as time passed interest faded, especially after Charles died and James went into exile. We shall never know for sure what happened, but the evidence collected and presented by Cook persuasively makes the case that the elder Halley's death was not a simple mugging gone wrong.

Whatever the cause of Edmond Halley senior's death, it changed the younger Halley's life dramatically, not least because his father had died intestate, and 'our' Edmond Halley was not on good terms with his stepmother. Flamsteed wrote to Molyneux in Dublin that:

Mr Halley I suppose may return to live in London, his father having been found drowned [sic] about a fortnight ago in Rochester river. This mishap will cause him a great deal of perplexed business, but nevertheless I hope to see him oftener than I have done of late.

If Flamsteed hoped to carry out joint observations with Halley, those hopes were dashed. After the events of March and April 1684 Halley virtually gave up observing, including his plan to chart the entire lunar cycle, until, eventually, he succeeded Flamsteed as Astronomer Royal.

Whatever his other problems following the death of his father,

Halley was certainly not hard up. It is quite likely that some real estate, probably including the house in Islington, had already been passed on to him while his father was alive, as was the usual practice, but we only know about the personal estate (money and other valuables, including leases on the rented properties) that was the subject of legal action and therefore a matter of record. Halley's father left £4,000, and the usual procedure was that a third should go to the widow (Joane), a third to be divided by the heirs (which meant it all went to Edmond, as his brother Humphrey was dead), and a third to be divided in accordance with the will. As there was no will, Joane tried to claim this third portion, and Halley fought her claim. When Joane remarried in 1685, further legal action resulted, with the eventual result that the property was divided roughly fifty–fifty. When the dust settled, Halley was left with an income from the various sources we know about amounting to £500–600 a year, with an unknown income from the other assets. He was always comfortably well off. It was in settling some of the financial affairs relating to the estate that Halley went to Alconbury in the summer of 1684, possibly also visiting relatives in Peterborough, and decided to call in on Newton in Cambridge on the way back to London. If not for the mysterious death of his father, that meeting would never have happened.

We have already described how that meeting led Newton to produce *De Motu*, and then, at Halley's urging, to begin work on what became the *Principia*. But while this was going on, Halley's situation changed again – once again in a way that would prove advantageous to the publication of Newton's masterwork – for reasons that are obscure, but seem to have been related to his changed circumstances following his father's death. We mentioned in passing earlier that Halley became Clerk of the Royal Society. But why? And how? The how is easier to answer than the why.

It was, as we have seen, part of the reorganisation of the Royal in the autumn of 1685, part of the aftermath of the Hevelius–Hooke controversy, which resulted in the resignation of the

Secretaries Aston and Robinson, and their replacement by John Hoskins and Thomas Gale. In January 1686, the Secretary of the Society laid out a set of rules for holders of the post of Clerk, which included the stipulations that he could not be a Fellow, must be unmarried and have no children, must live in Gresham College, and must have no other employment. But after going to the trouble of laying down these ground rules, the President, Samuel Pepys, announced to the Fellows that the choice of Clerk was entirely up to them, and that they could ignore any or all of those suggestions if they chose. They did choose. On 27 January, four candidates, including Halley, put themselves forward. Halley received sixteen votes, more than anyone else, but the others had twenty-two votes between them, so the lowest-ranked candidate dropped out and there was another round of voting. This time Halley received twenty-three votes out of thirty-eight and was elected as Clerk, being formally sworn in on 7 February. Just about the only one of the suggested conditions imposed on him was that he had to resign his Fellowship.

Why did he take on the job? It was certainly not, as has some-times been suggested, for the money. He had no need of the £50 annual salary that was offered by the Society. We say 'offered' because the Royal was very short of funds and seldom paid him, except in the form of copies of a book, *The History of Fishes*, which they notionally valued for this purpose at one pound each. The book had been published at considerable expense and had proved hard to get rid of, so they had plenty of copies to spare.* It seems that Halley was motivated by ambition: he wanted to be at the centre of things at the Royal, with his finger on the pulse of science, with an opportunity to influence events, make his own contribution and to be seen doing it. He certainly acted more like a secretary than a clerk, corresponding with scientists across Europe, editing and publishing the *Philosophical Transactions* at his own expense, and, of course, overseeing the publication of the *Principia*.

---

* It was a perfectly good book, but had been published in a lavish edition with many plates and was too expensive to attract many purchasers.

Halley seems to have had something of the same work ethic as Hooke. Although the post of Clerk was ostensibly a full-time job, he produced a steady stream of his own scientific contributions, both while seeing through the publication of the *Principia* (itself a project involving a lot of time and effort) and afterwards. In 1686 alone, he worked with Hooke to make observations of occultations of Jupiter by the Moon (part of the project to chart the changing position of the Moon), and produced three scientific papers of significance. One dealt with the relationship between barometric pressure, height above sea level, and changes in the weather. Another picked up on Hooke's ideas about the circulation of the atmosphere and developed an account of the trade winds, discussing the heating effect of the Sun on atmospheric circulation and the way the resulting pattern of winds was affected by the presence of land masses. The source of his interest in meteorology was mentioned in the paper, referring to his time on the island of St Helena: 'an employment that obliged me to regard more than ordinary the Weather.' As well as his own studies, he drew on many sources, including observations made by sailors and descriptions of the winds made by earlier writers, to produce what was in effect the first meteorological chart. The third paper was actually the first publication based on the ideas about gravity that Newton was about to present in the *Principia*, which Halley had already seen. It focused in particular on the application of the new understanding for gunnery, explaining how projectiles fired from cannon or a mortar moved under the influence of gravity, and how this information could be used to improve the accuracy of artillery. But it was for the publication of the *Principia* itself that Halley's early years as Clerk were most significant. The fact that he remained on friendly terms with both Newton and Hooke during and after this project is another testimony to his diplomatic skills.

Halley's formal involvement with the project began on 19 May 1686, when the Society passed the resolution that:

Mr Newton's *Philosophiae naturalis principia mathematica* be printed forthwith in quarto in a fair letter; and that a letter be written to him to signify the Society's resolution . . .

It was Halley, as Clerk, who wrote that letter, on 22 May, mentioning that 'I am intrusted to look after the printing it, and will take care that it shall be performed as well as possible'. But there was one other detail to be tidied up. The Royal still had no money, and on 2 June the Council 'ordered, that Mr. Newton's book be printed and that Mr. Halley undertake the business of looking after it, and printing it at his own charge; which he engaged to do.' So much for any suggestion that Halley needed his salary as Clerk; rather, the Society needed him to subsidise their work. In the event, Halley just about broke even on the project; he may have made either a small profit or a small loss, but nothing dramatic either way.

The plan was to publish the *Principia* in three sections, which were called 'books' even though they all appeared in a single volume within one set of covers. On 18 June 1686, Halley sent Newton a proof sheet of a few pages of the book, to get his approval of the paper and the type being used. It was in his reply to this letter that Newton wrote the intemperate screed of 20 June, mentioned in Chapter Seven, threatening to withold Book Three. 'Philosophie is such an impertinently litigious Laddy that a man had as good be engaged in Law suits as have to do with her,' he wrote. Halley did not act in haste but spent a couple of days mulling things over before composing a masterly response, dated 29 June 1686. He started by reassuring Newton that nobody wanted to take away the credit due to him:

I am heartily sorry, that in this matter, wherein all mankind ought to acknowledge their obligations to you, you should meet with anything that should give you disquiet, or that any disgust should make you think of desisting in your pretensions to a Lady, whose favours you have so much reason to boast of. Tis not shee but your Rivalls enviling

your happiness that endeavour to disturb your quiet enjoy-
ment, which when you consider, I hope you will see cause
to alter your former Resolution of suppressing your third
Book . . . These gentlemen of the Society to whom I have
communicated it, are very much troubled at it.

He then dismisses any suggestion that Newton should not be
regarded as the 'Inventor' of the proof of the inverse square law,
and moves smoothly on to business, pushing Newton into a
commitment by getting down to the nitty-gritty of publishing
details:

Sir, I must beg you, not to let your resentments run so high,
as to deprive us of your third book, wherein the application
of your Mathematicall doctrine to the theory of Comets,
and severall curious Experiments, which as I guess by what
you write, ought to compose it, will undoubtedly render it
acceptable to those that will call themselves philosophers
without Mathematicks, which are by much the greater
number. Now you approve of the Character and Paper, I
will push on the Edition Vigorously. I have sometimes
thought of having the Cutts [illustrations] neatly done in
Wood so as to stand in the page, with the demonstrations,
as it will be more convenient, and not much more charge,
if it please you to have it so, I will trie how well it can be
done, otherwise I will have them in somewhat larger size
than you have sent up.

Newton replied on 14 July, accepting the suggestion about wood-
cuts and, as we have seen, continued to work on Book Three as
Halley had urged. With the first two books in hand, by March
1687 Halley was employing not one but two printers to get it
published as soon as possible; apart from the importance of the
book, he may have been eager to get it out before Newton changed
his mind again. Book Three was delivered to him on 4 April, and
swiftly dispatched to the printers. To non-mathematicians this is

indeed the most important part of the *Principia*, and Halley was quick to highlight its importance in a letter to John Wallis, one of his old professors in Oxford. In Book Three, he says:

> is shown the principle by which all the Celestiall Motions are regulated, together with the reasons of the several inequalities of the Moons Motion and the cause and quantity of the progression of the Apogeon and the retrocession of the Nodes. How he falls in with Mr Hook, and makes the Earth of the shape of a compressed spheroid, whose shortest diameter is the Axis, and determines the excess of the radius of the Equator above the semiaxe 17 miles . . . he gives the reason for the tides . . . He concludes his book with the theory of Comets, showing that their Orbs are sufficiently near parabolicall; and upon that supposition shows how to find from observation the parabola wherein they move, and gives an exemplar of the Motions of the great Comet of 1680/1, and having stated the Orb from the observations of the Evening Comet, he finds that the comet that appeared in November in the Morning was the same . . .

The finished book was published on 5 July, Old Style. Halley received due credit in the preface:

> In the publication of this work the most acute and universally learned Mr. Edmund Halley not only assisted me in correcting the errors of the press and preparing the geometrical figures, but it was through his solicitations that it came to be published.

Halley's 'solicitations' did not stop there; he was actively involved in promoting the *Principia*, not least with a lengthy review in the *Philosophical Transactions*. This might seems today like a conflict of interests, since Halley was the editor and publisher of both the *Principia* and the *Transactions*, but it certainly did a great deal to spread the word to people lacking the mathematical sophistication to appreciate Newton's work first hand. He also gave a copy to the King, James II, somewhat exceeding his authority as Clerk (Pepys, on behalf of the Royal, should have made a presentation of the book in a more formal way), providing a masterly résumé of the main points of the book, pointing out that as a former naval person James would, Halley hoped, be particularly interested in the discussion of tides, and offering to go over anything with the King if 'by reason of the difficulty of the matter there be anything herein not sufficiently Explained'. This is another hint of Halley's ambitious nature, self-confidence and willingness for self-promotion, echoing his St Helena project. After 1687, however, with the *Principia*, including Book Three, safely off his hands, Halley was free to concentrate on his duties as Clerk, and his own scientific work.

Halley was nearly the equal of Hooke in the breadth of his scientific efforts, and did far more than Newton, who apart from his one piece of brilliance spent far too much time on pointless investigations of theology and alchemy. Halley's efforts as Clerk, and as editor of the *Philosophical Transactions*, were also something of a diversion from his own work, but of far more value to science than Newton's diversions; we shall not, however, dwell on them too much, since any competent administrator could have done the work. What interests us is Halley the scientist, and Halley the explorer.

Even in 1687, Halley published papers on geometry (again demonstrating his skill as a mathematician), on the evaporation of seawater, and one of the early scientific attempts to explain the biblical Flood. Hooke had suggested that the Flood might have been linked with a wobble of the Earth, which brought the region affected by the Flood under the waters bulging out at the equator. Halley was able to test this idea by comparing astronomical observations made at Nuremberg some centuries earlier with observations made in the seventeenth century. This showed that the latitude of Nuremberg had not changed in historical times. Halley offered an alternative explanation: that the Deluge had been caused by the gravitational influence of a comet passing close by the Earth, which would also have produced 'a change in the length of the year, and the eccentricity of the earth's orbit, for which we have no sort of authority'. The idea was wrong, because, as we now know, comets are nowhere near massive enough to have such an effect. But it was a sensible scientific suggestion at the time, and it carried the implication that whatever the physical events leading to the Flood were, they happened much longer ago than the time allowed by the timescale then favoured by the Church. Halley did not spell that out in 1687, but developed his ideas privately, and in 1724 published a paper in which he carried out a detailed analysis of the story in Genesis, where he suggested that the origins of the Flood story were 'lost by length of time' and that there must have been 'a much fuller account of the Flood left by the patriarchs to their posterity'. In other words, the story of the Deluge originated from an ancient verbal tradition, stories passed down the generations since times long past, and was not a divine revelation to Moses. This very modern approach to biblical stories was almost heretical, even in 1724, and demonstrates Halley's rational thinking.

On the experimental side, Halley carried out tests to try to determine how the strength of the force of magnetic attraction falls off with distance, but without success. In 1688 he carried out a long investigation of heat, evaporation, the causes of rainfall, and the design of thermometers. He described to the Society a type of fern that he had found on St Helena, and discussed the need of sunlight for green plants to thrive. But alongside the variety of his other activities, Halley was always interested in measurements, particularly astronomical measurements. As well as his own observations, he was involved in the publication of ephemerides (tables of the predicted positions of astronomical objects) in his capacity as Clerk. And, of course, there was always the problem of finding longitude at sea at the back of his mind. By the end of the 1680s, however, Halley and Flamsteed, the other great observer of the skies, had fallen out, and were not on speaking terms. The exact reasons are not clear, but Flamsteed seems to have regarded Halley as an upstart who had designs on his post as Astronomer Royal (quite possibly true). Flamsteed regarded his observations as his personal property, to be hoarded and used as he thought fit, which would eventually lead him into confrontation with the Royal; Halley regarded science as an open book, and was happy to make use of other people's observations, which Flamsteed saw as theft, or plagiarism. Halley was equally happy for other people to make use of his ideas. Flamsteed, who was a prim and puritanical man, seems also to have taken exception to some aspect of Halley's private life, which is not otherwise recorded. In 1695, in a letter to Newton, he referred to Halley as having 'almost ruined himself by his indiscreet behaviour' and alludes to deeds 'too foule and large for [discussion in] a letter'. Tantalisingly, we have no clue as to what those deeds were.

By the end of the 1680s, Halley was also involved in maritime projects of one kind or another. Exactly when and how these started we do not know, but on 22 March 1689 Hooke mentions in his diary 'Hally a sayling' and on 3 April 'Hally returned'.*
Rather than discussing these activities piecemeal, we will describe all Halley's work at sea, or all that there is a record of, in Chapter Ten. But while we concentrate on the science, keep in mind that there was this other string to his bow, enough for a full career for most men.

Something for which Halley ought to be remembered, but is seldom mentioned, is that he was the first person to make a scientific estimate of the size of atoms. In the 1690s, the possible existence of atoms was a matter for philosophical debate rather than practical science, but Halley, with his fascination for numbers and statistics, saw a way to cut through the debate. Silver-gilt wire, he knew, was made by drawing out a silver wire from an ingot which had a layer of gold around its circumference. In 1691, he asked gilders how much gold was used in the process, how thick the ingots were, and how thin the layer of gold was. From

---

* This was around the time of the 'Glorious Revolution'. Halley avoided politics, and is on record as saying 'For my part, I am for the King in possession. If I am protected, I am content. I am sure we pay dear enough for our Protection & why should we not have the Benefit of it?'

this information he calculated that the thin layer of gold on the final wire could be no more than 1/134,500th of an inch thick. This layer had to be at least one atom of gold deep, and he worked out that on this basis a cube of gold with sides one-hundredth of an inch long contains at least 2,433 million atoms. This would imply a maximum diameter for each atom of about five millionths of an inch, or 120 nanometers. Halley fully appreciated that in reality the atoms must be even smaller than this, and wrote that they were 'probably many times lesser, if the united surface of the Gold without Pores of Interstices be considered' (the modern figure is 0.14 nm). This was a real scientific measurement, made at a time when very few people took the idea of atoms seriously.

A couple of years later, Halley picked up on the demographic studies that he had begun with his analysis of births and deaths in Paris during his Grand Tour. He obtained similar data for the city then known as Breslau (now Wrocław). He was particularly interested in these data, because as far as he knew the population of Breslau was steady; the equivalent figures for London were distorted because the population was rising as people moved into the city. By looking at how many people of each age died in an interval of five years he was able to work out, among other things, the probability that somebody of a certain age would live on for a certain number of years. This kind of calculation produces actuarial tables that have been used as a basis for calculating life insurance premiums and annuities ever since. In one of Halley's own examples, he showed that for this particular population the cost of an annuity at a rate of 6 per cent for a man aged forty should be 10.57 years' premium. Some would live longer than average and do well out of such an annuity; some would die sooner and receive less. But on average the insurance company would balance its books. It was Halley who put life insurance on a scientific basis.*

---

* The system is sometimes distorted today by charging higher premiums for 'high-risk' individuals. The whole point of the actuarial approach is that the tables should be based on the entire population, not cherry-picked. But cherry-picking makes more money for the insurance companies.

In September 1691, in a presentation to the Royal Society, Halley discussed the possibility of measuring the distance to the Sun using a transit of Venus, but at that time he did not follow this up with a detailed analysis. At about that time, in spite of his continuing interest in astronomy and demonstrated skill as an astronomer, Halley suffered a setback in his astronomical career. That year, the post of Savilian Professor of Astronomy at Oxford became vacant, and Halley applied for the job. In purely scientific terms, he was an ideal candidate, but other considerations intervened, and there was also another worthy candidate.

The suitability of Halley for the post had been recognised in some quarters as early as 1678, when his career was taking off on the strength of the trip to St Helena. In that year, when Edward Bernard, who held the chair at the time, mentioned to Flamsteed that he was thinking of resigning, Flamsteed had recommended Halley as his successor, writing:

He is very ingenious, as I found when he talked with me; and his friends being wealthy, you may expect that advantage [that is, financial reward] by a resignation to him . . .

But by the time Bernard actually did resign, in 1691, a lot of water had flowed under the bridge, not least concerning Flamsteed's views about Halley. In the summer of 1691, Halley was busy with attempts to salvage the cargo from a ship, the *Guynie*, that had sunk off the Sussex coast. This threatened to interfere with his application for the professorship, and on 22 June, in a letter to Abraham Hill, one of the governors of the company that owned the ship, Halley not only brought him up to date with the salvage work but went on:

This business requiring my assistance, when an affair of great consequence to myself calls me to London, viz, looking after the Astronomy-Professor's place in Oxford, I humbly beg of you to intercede for me with the Archbishop Dr. Tillotson, to defer the election for some short time 'till I have done here, if it be but for a fortnight: but it must be done with expedition lest it be too late for me to speak. This time will give me an opportunity to clear myself in another matter, there being a caveat entered against me, till I can show that I am not guilty of asserting the eternity of the world.

The Archbishop was one of the Electors for the Savilian Chair, and it was essential that a candidate should be of orthodox religious views (or at least be thought to be; Newton got away with his unorthodoxy by keeping it secret); Halley's modest suggestions about the Flood seemed to be about to cause problems.

In the event, the election was not held until December, because Bernard did not get around to formally resigning his post until the beginning of November 1691. This gave Halley time to get an endorsement from the Royal Society, where at the meeting of 11 November it was his no doubt pleasant duty as Clerk to record that:

It was ordered that the Society doe give a recommendatory Letter to Mr Halley signifying their opinion of his abilities to perform the Office of Professor of Astronomy in Oxford now vacant, as likewise to testifye, what he has done for the advancement of the said Science, and that Dr Gale be desired to draw up the Testimoniall.

It was to no avail, and nor were Halley's protestations that he did not claim that the age of the Earth was infinite. The post went to the Scot David Gregory (not to be confused with James, inventor of the Gregorian telescope) who, ironically, had been ejected from his position in Edinburgh on religious grounds. There are no contemporary accounts of exactly what happened at the election, only later stories based on half-truth and rumour. There isn't even any direct evidence that Halley's ideas about the age of the Earth were held against him. But it is clear that one powerful voice in particular spoke up in opposition to Halley being appointed. On a letter he received from Wallis at the end of 1698, which referred to some work by Gregory, Flamsteed wrote:

> Dr Gregory is a freind of Mr Halleys tho he was his compet-
> itor but I perceive by this transaction he is no freind of mine
> tho I showed him more freindship than he could reasonably
> expect on yt occasion & Mr Halley as much enmity . . .

Quite a turnaround from 1678! Regardless of Flamsteed's animosity and any other caveats, though, Gregory was an excellent candidate who thoroughly deserved the post, so the result of the election should be regarded as his success rather than as Halley's failure. But the 'failure', such as it was, meant that Halley was available for another unusual career move in the 1690s. First, though, he made more contributions to astronomy, including his ground-breaking study of comets.

Before he carried out his comet studies, Halley made another astronomical investigation, which demonstrates both the breadth of his own knowledge and his willingness to embrace new ideas. He had a long-standing interest in ancient history, and found that this interest combined with his interest in astronomy in the work of the Arab al-Battani, also known as Albategnius, who lived from about 858 AD to 929 AD. Al-Battani had made observations of the Moon, which had come down to Halley's day in a Latin translation made by Plato Tiburtinus. Halley commented that, judging by this translation, Tiburtinus knew neither astronomy nor Arabic, and Halley used his knowledge of both to revise the text and draw out the correct astronomical information. Halley was something of a linguist, as this example shows. He studied Latin and Greek at school, knew some Hebrew as well as Arabic, and wrote letters in French and Italian while he was Clerk (as well, of course, as in Latin) and on other occasions. Contemporaries also said that he spoke German. With the correct data, Halley was able to work out the latitudes of al-Battani's observing sites, and then to compare his observations from those sites with seventeenth-century observations from London. What he found was dramatic.

As well as his routine observations of the Moon and other objects, al-Battani had observed solar eclipses in 891 AD and 901

AD, and lunar eclipses in 883 AD and 901 AD. The precise timing of all these events, combined with his 'modern' data, showed Halley that in the eight centuries between the Arab observations and his own the Moon had been speeding up in its orbit. We now know that this is a result of tidal influences: the rotation of the Earth slows down, while the Moon accelerates in its orbit. But the details of the physics do not matter here. What matters is that Halley had observed that the heavens are changing, and changing on a timescale noticeable to humankind. Hooke, as we describe shortly, had shown that the Earth evolves; Halley discovered that the heavens evolve. It was a profound, even revolutionary, idea that was essentially ignored at the time, and which gets only a passing mention in most books about Halley today.

In 1692, Halley published his ideas about terrestrial magnetism, which he had been mulling over, off and on, for some time. He suggested that the way the Earth's magnetic field varied could be explained if the interior of the planet was made up of an outer shell and an inner core, separated by a fluid layer, so that they could rotate at different rates. On this picture, each of the two parts of the Earth had its own magnetic field, with north and south magnetic poles, so there would be four magnetic poles in all, and the variations would be explained by the differential rotation of the two parts of the planet changing the geographical relationship of the two sets of magnetic poles. It was, he said, an 'Hypothesis which after Ages may examine, amend or refute.' It did, indeed, turn out to be wrong. But it was a reasonable and genuinely scientific attempt to explain a natural phenomenon, using the still-new idea of offering hypotheses based on past observations, which could be tested by further observations and experiments.

Halley was also, like Hooke, happy to apply his intellectual powers to more down-to-earth matters. Around the same time that he was investigating the orbit of the Moon, he was asked by one of the Fellows, James Houghton, how to work out the acreage of land in each English county. Halley's ingenious solution is an early example of lateral thinking. He took a large map of England,

and cut out the largest complete circle he could from the map. The circle covered a diameter representing 69⅓ miles, with an area corresponding to 9,665,00 acres. He then weighed both the circle and the complete map. Since the circle weighed only one quarter as much as the whole map, he concluded that the area of England was 36,660,00 acres. This differs by only a little over 1 per cent from the modern figure. He then worked out the area of each county by cutting their outlines out and weighing them.

Halley solved that problem without using any mathematical skill beyond the simple arithmetic he learned in school. But when he turned his attention to comets in the mid-1690s, he used the most sophisticated mathematical toolkit of the time, the technique, now known as calculus, which Newton had invented.* Halley was, indeed, the first person apart from Newton to apply calculus to the study of the Universe.

Newton himself had worked out the behaviour of things discussed in the *Principia* using calculus, but had not yet revealed the technique and knew that his contemporaries would not understand the calculations. So he recast all the calculations as geometrical examples, which was itself something of a mathematical *tour de force*. But Halley does not seem to have used geometrical methods at all in his attack on comets, since he wrote to Newton as early as 21 October 1695 that he was 'ready at the finding a Cometts orb by Calculation' – the calculation referred to being the techniques of calculus.

Although Halley began his detailed study of comets in the mid-1690s, and corresponded with Newton about his results, he continued to work on the problem intermittently over the next ten years, interrupted by other activities, and published most of his results in the *Philosophical Transactions* in 1705; the final version, with some relatively minor further tweaks, appeared in 1726. But we shall describe all of the comet work together, rather than scattering it piecemeal through the rest of our book. The

---

* *Calculus* was discovered independently by the German Gottfried Wilhelm Leibniz, which led to a bitter row about priority with the prickly Newton.

calculus technique was essential for calculating the orbits of comets accurately, because of the need to extrapolate from the relatively few observations that could be made while the comet was near to the Sun, and bright enough to be visible. The key question that Halley set out to answer was whether any comets are on elliptical orbits, which might take them far out from the Sun, but which were closed, so that those comets would eventually (and predictably) return to the inner part of the Solar System. Such orbits are much more elongated versions of the orbits of the planets, which, although elliptical, are nearly circular. The main alternative was that comets might follow parabolic orbits. These are open, in the sense that a comet in such an orbit falls towards the Sun from the depths of space, swings round the Sun, and disappears off into the void, never to return. There was also the possibility that they might follow hyperbolic orbits, which are also open and from our point of view can be regarded as extreme parabolas. One of the spin-offs from Newton's proof of the relationship between orbits and the inverse square law was that it showed that these kinds of orbits are also allowed by the same inverse square law. This confirmed Hooke's suggestion that comets are ordinary members of the Solar System, subject to the same laws as the planets, and opened up the possibility of predicting their orbits.* The problem was that in the parts of these orbits close to the Sun, where comets are visible, all three kinds of orbit follow very similar paths, and look very similar. You would need very accurate observations and very detailed calculations to distinguish one kind of orbit from another. That was (just) what Halley had, although in the *Principia* Newton himself had concluded that cometary orbits were parabolic.

One reason for Halley's success was that he went back over the historical records and used every scrap of information he could lay his hands on, as well as the more or less contemporary seventeenth-century records from people such as Hevelius and

---

* It had only been in the 1570s, little more than a hundred years before the *Principia*, that Tycho Brahe had proved that comets travel past further away from us than the Moon is.

Flamsteed, and his own observations. He was lucky to have so much data: compared with twentieth-century observations, and the twenty-first century so far, an unusually large number of bright comets had been visible in the sixteenth and seventeenth centuries. Halley had enough observations to compute the orbits of twenty-four comets, each one, as he put it, an 'immense labour'. And two of the sets of calculations had to be redone when he found that some adjustments were necessary. One of these was for the comet of 1682, now known as Halley's Comet. Halley suspected that this was the same as a comet that had appeared in 1531 and one seen in 1607. If that were the case, it must be following an elliptical orbit. But when he first calculated the orbit, it came out as a parabola. He needed more data, and the way he got it shows something of the scientific politics of the time.

Flamsteed had made detailed observations of the comet, which were just what Halley needed, but which, characteristically, Flamsteed had not published. Flamsteed was not speaking to Halley, and there was no chance of getting the data directly from him. So Halley was forced to write to Newton:

> I must entreat you to procure for me of Mr Flamsteed what he has observed of the Comett of 1682 particularly in the month of September, for I am more and more confirmed that we have seen that Comett now three times, since ye Yeare 1531, he will not deny it you, though I know he will me.

Newton got the observations from Flamsteed, and passed them on to Halley, who re-computed the orbit and found that it was indeed elliptical, with a period of about seventy-six years. That made it possible, in principle, to predict exactly where the comet would appear on its next return. But it wouldn't be easy.

In order to predict exactly where on the sky the comet would reappear, Halley needed to calculate the gravitational influence of the giant planets Jupiter and Saturn on the comet when it passed through the outer Solar System, following Hooke's insight that all astronomical objects exert a gravitational influence. Even Halley needed help with this. He wrote to Newton:

> I must entreat you to consider how far a Comets motion may be disturbed by the Centers of Saturn and Jupiter, particularly in its ascent [return] from the Sun, and what difference they may cause in the time of the Revolution of the Comett in its so very Elliptick Orb.

Newton replied to the effect that the perturbing influences of the planets on a comet could not be calculated without a lot more information about the particular orbit. Neither Newton nor Halley ever worked out a satisfactory way to deal with the perturbation problem. Even with that aid, he would have had to work out where Jupiter and Saturn would be at the relevant time to get an approximate indication of their distance. Finally, having adjusted the orbital calculation for the comet accordingly, he would have had to work out where the Earth would be in its orbit at the appropriate time, so that he could tell astronomers on Earth where to point their telescopes to catch an early sight of the comet. It was left for Halley's successors* to predict, building on his work, not just that the comet of 1682 (and 1607 and 1531) would reappear in 1758, but that it would first be seen in a particular part of the sky around Christmas 1758. We shall return to this famous prediction, and its consequences, in Chapter Eleven. But now it is time to look at some of the distractions that kept Halley occupied and largely away from astronomy between 1696 and 1703.

Just a year after his letter to Newton claiming to be 'ready at the finding a Cometts orb', on 21 October 1696 Halley informed Newton that:

> I have almost finished the Comet of 1682 and the next you shall know whether that of 1607 were not the same, which I see more and more reason to suspect.

---

* Notably the Frenchman Alexis-Claude Clairaut.

Around the same time, in an undated letter, he wrote:

> I will waite on you at your lodgings to morrow morning to discourse the other matter of serving you as your Deputy.

Newton was by now, of course, Warden of the Royal Mint, and living in London. The post of 'Deputy', which Halley refers to, was an invitation, which he accepted, to take charge of the recoinage that Newton was overseeing at one of the regional mints, at Chester. It was not to prove a happy experience.

The recoinage was necessary because the practice of 'clipping' silver coins – snipping off small pieces to melt down and turn into forged currency – had become so widespread that the value of the currency was sliding and inflation was setting in. The solution was to call in all the old silver coins and replace them with new coins with milled edges that could not be clipped without this being obvious. The Chancellor of the Exchequer at the time was Charles Montagu (later the Earl of Halifax), a Fellow of the Royal Society and a friend of Newton, which explains Newton's appointment to the Mint in 1696. Halley was not actually Newton's deputy in Chester, although he was appointed at Newton's suggestion. In fact, he was Deputy Comptroller there, the Comptroller being one William Molyneux, who was responsible to the government.

Newton must have thought he was doing Halley a favour by finding the post for him, perhaps as a mark of gratitude for his work on the *Principia*, but it is not clear what the reason was, nor why Halley accepted the offer. One possibility is that it was a way to raise Halley's profile as a public servant and gain favour with the King, but this seems unlikely given Halley's already high reputation. Another suggestion is that Halley was temporarily short of money, and needed the income, which was £90 a year. In 1693, English and Dutch merchants had suffered a large financial loss when a fleet of ships was intercepted by the French off Lagos, in the south of Portugal. This became known as the Lagos disaster. Speculators who had invested in cargoes carried by the ships lost everything. Halley, who had contacts among those traders, seems to have been one of the investors; Hooke's diary entry for 24 July 1693 tells us '2 East India ships said to be taken by French in India. Hot, clear. Hallys trade taken by French.' The presence of the weather report in between the other two sentences suggests that Hooke is referring to two separate incidents, and that Halley's loss was not related to the ships taken by the French in India, but to the Lagos disaster, news of which reached London on 24 July. Or maybe (unlikely, in view of his meticulous record-keeping), Hooke confused two separate events. But would Halley have still been in financial need three years later? Surely he would have sought an income sooner if he had been badly hit by the Lagos disaster. We shall never know, but for whatever reason, Halley did take up the appointment, and managed to carry out his duties as Clerk at the same time.

Very little needs to be said here about Halley's time in Chester. The clerks there were impudent and unsupportive, and quarrelled among themselves to the point of one challenging another to a duel; although the challenge was accepted, the duel never actually took place. There was also, almost inevitably, the kind of financial irregularities (for which, read theft) that was all too easy when large quantities of silver in small denomination coins were being handled. On 25 October 1697, Halley wrote to Hans Sloane, one of the Secretaries at the Royal, apologising for being unable to attend a meeting of the Society:

> as yet the business of our Mint is not in such condition, that I can be spared for good and all, although in a month I guess wee shall have finished our whole coinage. . . my heart be with you and I long to be delivered from the uneasiness I suffer here by ill company in my business, which at least is but drudgery, but as we are in perpetual feuds is intollerable.

He only tolerated the situation a little longer, and resigned early in 1698, after an inquiry had looked into the abuses at Chester and found that they were no fault of his. It was almost time for Halley's greatest adventure, but first a brief interruption showed just how highly regarded he was in Court circles, with, surely, no need of Newton's patronage.

Halley's big adventure would involve voyaging to the South Seas on a ship, the *Paramore*, built especially for the job. As we describe in Chapter Ten, the project had been in hand for some time, but delayed by, among other things, the Nine Years' War with France (during which the Lagos disaster occurred), which was ended by the Peace of Ryswick, signed in October 1697. Early in January 1698, however, Czar Peter of Russia (later to be known as 'the Great') came to England to study shipbuilding techniques, and at Peter's specific request in March that year the *Paramore* was rigged and floated, not for Halley's expedition, but to be made available to be used 'as the Czar of Muscovy shall desire'.* Peter wanted to carry out sailing 'experiments' with the ship, and later said that he would 'far rather be an admiral in England than Tsar in Russia.'† This was just one of several ships made available to Peter, who was also given a free run of the dockyards at Deptford. The Czar was just twenty-five when he arrived in London, and very much a practical 'hands-on' man who was not afraid of getting himself dirty learning the practicalities of shipbuilding. He was also a hands-on man in other ways, with a reputation for womanising, heavy drinking and crude manners and language, not the sort of person Flamsteed would have approved of. Peter stayed at a house in Deptford which was owned by John Evelyn, but rented out at the time to Admiral John Benbow. By the time he left, on 21 April, the house and gardens had suffered so much damage that the Exchequer paid Evelyn £300 in compensation. He may have been short-changed; Christopher Wren estimated the cost of repairing the damage as nearly £400.

It is not known what relationship Halley had with the Czar,

---

* Admiralty letter dated 16 March 1698, reprinted in Thrower.

† I. Gray, 'Peter the Great in England', *History Today*, volume 6, 1956, pp 225–234.

but he seems to have been made available as an adviser on nautical (and possibly scientific) matters. It is likely that he sailed with Peter on some occasions, and certain that he dined at Deptford. There is no evidence whether or not he joined in the drunken escapades that caused so much damage, although it is hard to see how he could have refused a suggestion by the Czar to join in some fun and games.* In a biographical note by Halley's younger contemporary Martin Folkes we are told:†

> This Great Prince was highly pleas'd with him, treated him with great distinction, admitting him to the Familiarity of his Table, to have the more opportunitys of being inform'd by his [entertaining] and instructing Conversation.

And the *Biographia Britannica*, published in 1757, says that Peter:

> . . . found [Halley] equal to the great character he had heard of him. He asked him many questions concerning the fleet which he [Peter] intended to build, the sciences and arts which he wished to introduce into his dominions, and a thousand other subjects which his unbounded curiosity suggested; he was so well satisfied with Mr Halley's answers, and so pleased with his conversation, that he admitted him familiarly to his table, and ranked him among the number of his friends.

But however good his relationship with Czar Peter was, Halley must have been glad to see the back of him. After the tedium and troubles at Chester, he was about to experience a literal breath of fresh air, and to become, as Pepys described him, 'the first Englishman (and possibly any other) that had so much, or (it might be said) any competent degree (meeting in them) of the science and practice (both) of navigation.' His friend Robert

---

* One of Peter's favourite drunken games, apparently, involved being pushed at high speed in a wheelbarrow and crashing through the ornamental hedges in the garden. Three wheelbarrows were broken in this way during his stay.

† Reprinted by MacPike.

Hooke, himself a frustrated would-be traveller who took a keen interest in voyagers who had visited far-distant lands, must have felt a twinge of envy.

We can leave Halley getting ready to take the *Paramore* to sea while we recount what happened to Hooke after the publication of the *Principia*.

CHAPTER NINE

NOT FADE AWAY

In spite of what Richard Waller tells us, it is clear that Hooke did not fade away immediately after the death of Grace. In January 1687, his salary at the Royal had been increased £100 a year, half of it coming from the Royal itself and the rest (if he could be made to pay up) from Cutler. In return, Hooke was once again more active in the Society, carrying out experiments and providing written 'discourses' on the work.* Grace died in February that year, aged twenty-six, of 'a fever'. She was buried on the 28th, at the church of St Helen Bishopsgate. But Hooke's grief did not prevent him from completing a series of lectures on fossils that he had begun the previous December and ended in March. We shall save discussion of Hooke's ideas concerning fossils, the age of the Earth, and life for later in this chapter, after describing his other activities.

Socially, Hooke was as active as ever, enjoying the company of his friends in coffee shops (he still favoured Jonathan's), visiting

---

* It is not clear whether these payments continued after 1688, but by that time Hooke had no need of the income.

bookshops, and indulging his slightly extravagant taste in clothing (at the end of 1688 his new winter coat cost seventeen shillings and sixpence). He had, after all, less incentive, when he was not working, to stay in his rooms, where he was looked after by a maidservant, Martha. But when he took up diary keeping again the Pisces symbol did not appear anywhere alongside her name, or that of her successor. His frequent companion at home was Harry Hunt, now the Operator at the Royal, who often dined with Hooke, took tea with him and ran errands. The relationship has been described as like father and son.*

The diary begins again in November 1688, just after William of Orange had landed in England with his army, and gives us some insight into how the kind of people Hooke mingled with viewed the 'Glorious Revolution'. The invasion began at Torbay, in the West Country, but William moved slowly towards London, gathering support without bloodshed, while support for James declined to the point where the army abandoned him to his fate.

Hooke records some civil unrest in London during November, when it was not clear whether James would try to make a stand against the invader, and he writes on 5 December of 'Great confusion of reports'. The confusion was largely caused by James dithering. He eventually decided to make a run for it on 11 December, which triggered some rioting and looting. '[R]able rifled Salisbury house', Hooke wrote, and 'the tower surrendered to Lord Mayor and Lord Lucas Governor disarmed all Papists in the tower.' James was soon captured and brought back to London, but allowed to leave by ship from Greenwich on 18 December, the day William arrived in London. Hooke writes that a few weeks later in Jonathan's he 'pleaded against Division and Revenge'. He was not alone. The consensus seems to have been that even a Dutch king was preferable to another civil war. Halley, as we noted earlier, wrote at the time:

---

* 'Espinasse.

For my part, I am for the King in possession. If I am protected,
  I am content. I am sure we pay dear enough for our
  Protection, & why should we not have the Benefit of it?

And, indeed, life went on pretty much as normal in London
while the position of William and Mary was being formalised.

It is striking that at the end of the 1680s, with Hooke in his
mid-fifties, there is less mention in the diary of the vomiting and
purges that feature in the earlier volume, and Hooke seems to
have been in good health (by his standards), and fit enough for
long walks, although Richard Waller tells us that earlier in 1688,
before the diary resumed, Hooke had been ill with 'Head-achs,
Giddiness and Fainting, and with a general decay all over, which
hinder'd his Philosophical Studies'. Although this seems to have
passed, he now had increasing trouble with his eyesight, which
had probably been damaged by the strain of his microscopic work,
and this may explain brief references in the diary to feeling melancholy. He served on the Council of the Royal every year from
1689 to 1695, in 1697, 1698 and 1700, and took what opportunities he could to establish his priority over ideas that were now
being rediscovered by others.

In a lecture to the Royal on 26 June 1689, he deviated from
his subject (the mixing of liquids) to complain:

Though many of the things I have first Discovered could
  not find acceptance [at the time] yet I find there are not
  wanting some who pride themselves on arrogating of them
  for their own – But I let that passe for the present.

He had a point. Early in 1690, for example, Halley came up
with the idea of using telescopic sights in quadrants, and Hooke
grumbled to his friends at Jonathan's that 'Hally pretended to the
glasse sights of Sea quadrant, though it was printed in *Hist. of
the Royal Society* before he went to school'. That, of course is
why Halley knew nothing of it – he had not been reading such
publications when a schoolboy, and didn't know of Hooke's work.

But he soon acknowledged it, and the incident did not harm their friendship.

In another presentation in February 1690, Hooke highlighted another piece of Newtonian plagiarism, sarcastically saying that 'of late Mr. Newton has done me the favour to print and publish as his own Inventions' several of Hooke's ideas, including the equatorial bulging of the Earth:

> And I conceive there are some present that may very well Remember and Doe know that Mr Newton did not send up that addition to his book till some weeks after I had read & shewn the experiments and demonstrations thereof in this place.

But although Hooke had a point, protestations like these helped to create an image of a cantankerous old man critical of the work of others. That image has unjustly coloured much of the writing about Hooke since his death, including discussions of his earlier life and work.

A year before Hooke's mutterings about Halley, Halley himself brought Newton, in London for the Convention Parliament, and Hooke together at his house; they met again at the Royal on 12 June 1689, when Huygens and Fatio were also present. Their paths crossed several times that year, but since these meetings went largely unremarked it seems that there was neither a reconciliation nor a major confrontation. What is more telling is that Newton had very little to do with the Royal, even after he moved permanently to London, while Hooke was still alive. Hooke's presence was clearly the reason for this, since following Hooke's death Newton became *the* major figure in the Society. As for Hooke, although peeved that Newton had stolen his ideas, he had plenty of other things to do, and doesn't seem to have let the dispute affect him drastically. Unlike Newton, he was not obsessive, and continued with a wide variety of activities, including visiting the ship of his friend Captain Robert Knox back from his latest travels, watching an eclipse of the Moon, architectural work (his friendship with Wren continued), microscope design, and much more.

One of the most intriguing stories brought back by Knox from India concerned 'a strange intoxicating herb', like hemp, known as gange. The self-dosing Hooke was thoroughly intrigued by Knox's first-hand account of experimenting with the drug, but never managed to grow it and try it himself. He described the effects, as reported by Knox, to the Royal:

> [the patient] is very merry, and laughs, and sings, and speaks Words without any Coherence, not knowing what he saith or doth; yet he is not giddy, or drunk, but walks and dances and sheweth many odd Tricks; after a little Time he falls asleep, and sleepeth very soundly and quietly; and when he wakes, he finds himself mightily refresh'd, and exceeding hungry.

Although he never did get to go on a long voyage, Hooke always took advantage of any opportunity for a short sea trip, usually to test his navigational instruments, and often joined in excursions along the river, like the outing of May Day 1665, which is described in Pepys' diary. On 22 March 1689, Hooke's diary notes 'Hally a sailing'; on 3 April he notes 'Hally Returned'. Reading between the lines, we can guess that Hooke would have welcomed an invitation to join Halley on this mission to survey the Thames approaches.

There is a gap in the diary – a missing volume – covering March 1690 to December 1692, an interval which saw the death of two old frends, Theodore Haak in May 1690 and Robert Boyle in December 1691, plus the award to Hooke of the degree of Doctor of Physick (Medicine) by the University of Oxford, on the advice (meaning order) of the Archbishop of Canterbury, John Tillotson, 'as he is a Person of a prodigious inventive Head, so of great Virtue & goodness; and as exceedingly well vers'd in all mathematical & mechanical, so particularly in astronomical knowledge'. Microscopy also remained linked with Hooke; in his will, Boyle left Hooke 'my best microscope and my best loadstone.' Hooke's ongoing work as surveyor and architect included supervising various repairs and improvements to Westminster Abbey, and the design of several buildings at Plymouth Docks, including housing for officers. Pevsner describes this as 'the first example in any of the English dockyards of a unified approach for officers' housing.'* He was also still involved in designing better instruments for navigation. It was in this connection that he made a perceptive remark in one of his lectures at the end of 1690:

> There are many things, that before theory are discover'd, are look'd upon as impossible, which yet, when they are found, are said to be known by every one, the inventor only excepted, who must pass for an Ignoramus.

This has a pre-echo of the famous statement by the German philosopher Arthur Schopenhauer:

> All truth passes through three stages. First, it is ridiculed. Second, it is violently opposed. Third, it is accepted as being self-evident.

When the diary entries resume in December 1692, Hooke, now fifty-seven, is on the Council of the Royal, engaged in several building works, and still fit enough to travel everywhere around London on foot. But for him, original science was now essentially a thing of the past. His interest in travellers' tales of the world beyond Europe remained strong, and right up until the final entry on 8 August 1693 we have an image of the same old Hooke, far from being 'Melancholly and Cynical', still frequenting coffee houses, and still able, in spite of trouble with his eyes, to read the books he continued to purchase in great quantities. But he was not as physically strong as he had been, unable to clamber over scaffolding, and with deteriorating eyesight. He formally

---

* *The Buildings of England, Devon*, 1989.

ceased his activities as Wren's partner in 1693. But his mind was as sharp as ever. Knox, back in London in the spring of 1694, told Hooke, among other things, about the giant leaves of the taliput tree, which provided welcome shade in tropical climes. In the 1690s, it was widely accepted that all things on Earth had been put there by God for the benefit of mankind, so these trees must have been created for the purpose of providing people with shade. But Hooke told the Royal that the leaves were there for the good of the tree, and the fact that they happened to provide shade for humans was a coincidence. Obvious to us, but not to the Fellows of the time.

It was his interest in tales from far lands that led Hooke to design and build a 'Picture-Box', a kind of portable camera obscura that would project an image of the outside world on to paper so that it could be sketched quickly and (more importantly) accurately.

The tedious wrangle over payment for the Cutlerian lectures also came to a denouement in the mid-1690s. With Cutler dead, Hooke (supported by the Royal) had to battle through the courts with Cutler's executors, finally winning his case on his own sixty-first birthday, 18 July 1696, with his arrears back to 1684 paid and an order that the estate should continue to pay £50 a year as long as he continued to give the lectures. The money was less important to him than the principle, but he continued to give the lectures until May 1697, by which time his health was failing. In the summer of that year, he began to suffer with sore and swollen legs, reducing his mobility, and had a fall that seems to have cracked some of his ribs. Waller tells us that by that time he had ceased to eat meat, 'no flesh in the least agreeing with his weak constitution', and subsisted on a diet of milk and vegetables. The Royal Society still met at Gresham College, so he was able to attend meetings, give lectures (most notably on earthquakes) and serve on Council. His last contribution was on 24 June 1702, when he made a short comment about earthquakes and glass-grinding.

It was in these declining years that Hooke began to neglect himself, and become over-concerned that he might not have enough money to see out his life. As Knox put it, he 'lived Miserably as if he had not sufficient to afford him food & Rayment'. In Jardine's words, 'by the age of 65 Hooke was a physical wreck, emaciated and haggard', a situation she blames in no small measure on 'his self-dosing with pharmaceuticals':*

It is probable that Hooke's regimen was ultimately fatal, and that even before it killed him the side-effects of his medicines (clouded vision, giddiness, lassitude, melancholy) proved damaging and disabling.†

Less dramatically, others have argued that the symptoms described by Waller suggest a congestive heart failure, with the circulation of the blood becoming so weak that the extremity of the body, in particular the legs, 'swells with retention of fluid, constricting circulation even further.'‡ In any case, in his final year Hooke was all but bedridden, attended by a maidservant, and suffering breathing difficulties as well as the swelling of his legs. The end came in March 1703, in his sixty-ninth year, as reported by Knox:§

The 2th March 1702/3 This night about 11 or 12 of the Clock my Esteemed Friend Dr Robert Hooke Professor of Geometry & Naturall Philosophy in Gresham College Died there, onely present a Girle that wayted one him who by his order (just before he died) came to my Lodging & called me. I went with her to the Colledge where, with Mr Hunt the Repository Keeper, we layed out his body in his Cloaths, Goune & Shooes as he Died, & sealed up all the Doores of his apartment with my Seale & so left them.

Hooke was buried with due ceremony at St Helens (near Tom Giles and Grace Hooke); as Waller put it, 'decently and handsomely

---

* See her contribution to *London's Leonardo*.
† *Curious Life*.
‡ Drake.
§ See *Concerning Severall Remarkable Passages of my Life*, Glasgow, 1911.

interr'd in the Church of St. *Hellen* in *London*, all the members of the Royal Society then in Town attending his Body to the Grave, paying the Respect due to his extraordinary Merit'. But the location of his remains and those of his kin were lost during rebuilding work in the nineteenth century. There is, however, a stylised representation of him (far from true to life!) in the stained-glass window at the west end of the church. Hooke left no will, in spite of having talked of making a bequest to the Royal, and his material inheritance became the subject of an unseemly quarrel among his distant relatives, which need not concern us here. But there was plenty to quarrel about – more than £8,000 in cash, plus some gold, jewellery, and valuable books, as well as the usual personal effects.

Hooke's real legacy, of course, was his contribution to science. So rather than concluding the story of his life with its rather depressing end, we will summarise one of his lifelong scientific themes, which also happened to be the last topic of major scientific significance with which he addressed the Royal Society: the nature of fossils, the age of the Earth, and the evolution of life.

The definitive study of Hooke as a geologist has been carried out by Ellen Tan Drake, of Oregon State University, and published, along with Hooke's lectures on 'earthquakes' (a term he used to embrace all changes in the surface of the Earth including volcanic eruptions) in her authoritative book *Restless Genius*. We follow her presentation here, and the quotes from Hooke are taken from the transcriptions included in her book, which provides the most easily accessible and complete compilation of what Drake refers to as Hooke's 'earthly thoughts'.

Hooke gave several series of lectures on 'earthquakes' over an interval of more than twenty years. There is a considerable amount of repetition and overlap in the lectures, and some digressions where Hooke cannot resist wandering off the subject to discuss some other phenomenon. So we have selected quotes from different lectures, not always following the chronological sequence in which the talks were given, in order to make Hooke's thinking clear. But his main themes are indeed clear, and astonishingly modern.* It is also clear, as Drake has spelled out, that Hooke's thoughts on geophysics, as it is now called, were a direct influence on later Earth scientists, including James Hutton, who is widely regarded as the father of geology.

As we have mentioned, Hooke's boyhood on the Isle of Wight was of fundamental importance to his interest in geology, and he often used examples from the island to illustrate his ideas, at least as early as in *Micrographia*. In that book, he provided the first drawing of a foraminifer, and described how he had found it among grains of sand, as 'an exceedingly small white spot, no bigger than the point of a Pin', which resembled 'the shell of a Water-Snail, with a flat spiral Shell.' Nobody had previously noticed the existence of such microorganisms. In one of his lectures, he refers to a stone which he 'broke with a smart stroke of a Hammer' with the result that he 'discover'd two small Snake-stones within it, which probably had been tumbled into the Mouth of it before it was concreted; for they were of the very same Substance, but of a differing Figure from any I have yet describ'd'.

Hooke explained the existence of fossils in terms of eleven 'Propositions', starting with the assertion that such objects were indeed the petrified remains of living things, with 'their Pores fill'd up with some petrifying liquid Substance, whereby their Parts are, as it were, lock'd up and cemented together in their Natural Position and contexture', and ending with the statement that 'there have been many other Species of Creatures in former Ages' (more of this shortly). In support of the first proposition, Hooke describes the way stalactites and stalagmites form in caves solely by the action of dripping water, 'by which we are assured that Nature really does change Water into Stone'.

And he gives short shrift to those who subscribe to the then popular notion, described by Hooke as 'Astrological and Magical

---

* One example of his 'modern' thinking is that he appreciated the value of a negative experimental result. 'The discovery of a Negative is one way of restraint and limiting an Affirmative.' Or, in the words Arthur Conan Doyle put into the mouth of Sherlock Holmes, 'when you have eliminated the impossible, whatever remains, however improbable, must be the truth'.

Fancy', that fossils are just stones that resemble living things, people who think that:

> They were produced from some extraordinary Celestial Influence, and that the Aspects avid Positions of the fix'd Stars and Planets conduc'd to their Generation, so that they also have in them a secret Vertue whereby they do at a distance work Miracles on things of the like Shape.

Such ideas, he says, are 'fantastical and groundless'. He was certainly no alchemist.

Hooke was able to make the intellectual leap from the study of such remains to genuinely scientific generalisations about the long history of the Earth. And he realised that the story of Creation told in the Bible should not be seen as literally true – unlike Newton, for example, who in a desperate attempt to reconcile observations with the story of Genesis suggested that the Earth might have rotated more slowly in the past, so that the six 'days' of Creation occupied a long span of time. In December 1680, he wrote to Bishop Thomas Burnet, author of *The Sacred Theory of the Earth*,

> . . . all this might ye rather bee, because at first we may suppose ye diurnal revolutions of ye Earth to have been very slow, soe yt ye first 6 revolutions or days might containe time enough for ye whole Creation . . .

It doesn't seem to have occurred to Newton to wonder what the effect of the long nights would have been on the life that existed before the job of Creation was completed with Adam and Eve. Hooke (who, by the way, realised that the rotation of the Earth was slowing down, and must have been faster in the past) had a simpler explanation. He noted that the story in the Bible was written down long after the events it purported to describe, and had previously been passed down over many generations by word of mouth, providing ample opportunity for the story to get distorted. As he put it, before the invention of writing the stories must have been 'dark and confused', and 'cannot be much relied on or heeded'. Hooke was a religious man, in that he believed in God, but he was quite prepared to adjust religious ideas to fit the facts, rather than the other way around.

One of the facts that had to be explained was the presence of shellfish fossils far from the sea – indeed, far above sea level. 'Who can imagine that Oysters, Muscles, and Periwinkles, and the like Shell-fish, should ever have had their Habitation on the tops of the Mountain?' And not just on mountains. Fossils are found, Hooke points out:

> At the tops of some of the highest Hills, and in the bottom
> of some of the deepest Mines, in the midst of Mountains
> and Quarries of Stone, &c.

The reason for their presence there must be 'the prodigious Effects that have been produced by Earthquakes on the superficial Parts of the Earth'. But he does not suggest that these were violent events that raised mountain ranges up in a single catastrophe. Rather, he refers to changes in the surface of the Earth happening bit by bit ('by degrees'), gradually over, by implication, a long (very long) interval of time. 'Those prodigious Piles of Mountains are nothing but the effects of some great Earthquakes.' As an example, he reports descriptions of the eruption of the volcanic island of Santorini in 1650, and comments:

> Though our Natural Historians have been very scarce in the
> World, and consequently such Histories are very few; yet
> there has been no Age wherein such historians have liv'd,
> but has afforded them an Example of such effects of
> Earthquakes.* And I doubt not, but had the World been
> always furnisht with such Historians as had been inquisitive

---

* For Hooke, the term 'earthquake' includes volcanoes. He also refers to the craters of the Moon as having been formed by 'eruption . . . somewhat Analogous to our Earthquakes'.

and knowing, we should have found not only *Thera* or *Santerinum*, and *Volcano* and *Delos*, and that in the *Azores*, and one lately in the *Canaries*, but a very great part of the Islands of the whole World to have been rais'd out of the sea.

And he says:

Nor do I conceive they were all thus formed at once, but rather successively, some in one, some in other Ages of the World, which may probably be in some measure collected from the quantity or thickness of the Soil or Mould upon them fit for Vegetation.

This ties in with a broader conclusion he reached:

that a great Part of the Surface of the Earth has been since the Creation transform'd, and made of another Nature: that is, many Parts which have been Sea are now Land, and others that have been Land are now Sea; many of the Mountains have been Vales, and the Vales Mountains.

Hooke also notes that 'Convulsions', as he calls them, are most common in mountainous regions, and cites Pliny, who 'says, that the *Alps*, and *Appennine* Mountains have very often been troubled with Earthquakes'.

All of this ties in with another key feature of Hooke's thinking about the Earth. Although Copernicus had removed the Earth from its status as the centre of the Universe, it still seemed, especially in the light of the biblical stories, to be a place of special significance in the Universe. Hooke was the first scientist to treat the Earth simply as a planet, one member of the Sun's family of planets, distinguished only by our presence on its surface, but subject to exactly the same laws (such as gravity) as the other planets. Newton, influenced by Hooke, only slowly came to this realisation.

This image of the Earth in space contributed to Hooke's ideas about polar wandering, which involves a shift in the position of the poles as seen from the surface of the Earth, or, if you prefer, a shift in the crust of the Earth relative to the poles. This is not continental drift – the continents do not move relative to one another – but involves the solid crust of the Earth moving over a more fluid interior. But it is a real effect, which is an integral part of the modern theory of plate tectonics, and has actually been measured using geological data – it occurs at a rate of about 5–10 cm per year, and at present the North Pole is moving (from our perspective) towards London. Hooke's version invoked changes in the position of the centre of mass of the Earth, linked to geological processes. The motivation for this idea was the evidence Hooke saw from his fossil studies, including remains of tropical species found at Portland, that England had once been in what he called the 'Torrid Zone' and must have moved to its present position. 'Consider, whether it may not have been possible, that this very Land of *England* and *Portland*, did, at a certain time for some Ages past, lie within the *Torrid Zone*.' The package of ideas included in this concept included the Earth being more dense in the centre than at the surface (explaining why heavy metals such as gold are rare at the surface), an equatorial bulge produced by the rotation of the Earth, and the suggestion that 'whenever an Earthquake raises up a great part of the Earth in one place it suffers another to sink in another place'. Mineral ores, he suggested, were generated deep within the Earth and had been uplifted 'by some former Subterraneous Eruption (by which those Hills and Mountains have been made)'. While others – even the famous Steno, of whom more shortly – thought that fossil remains found on mountain tops were a result of Noah's flood, Hooke was unique among his contemporaries in suggesting that they had been lifted up by geological activity, so that:

the superficial Parts of the Earth have been very much changed since the beginning, that the tops of mountains have been under the Water, and consequently also, that divers parts of the bottom of the Sea have been heretofore Mountains.

As for Noah's flood, it might have happened, but it did not last long enough (a mere 200 days) to explain how the remains found in, for example, the Alps, had got there. Once again, biblical stories give way to scientific facts.

It's worth elaborating on Hooke's ideas concerning the shape of the Earth, since Isaac Newton is still often credited with the 'discovery' of its oblate shape. Hooke lectured on the subject in the mid-1670s, some twelve years before Newton published his version in the *Principia*. Newton's correspondence with Burnet shows that at least as late as 1681 he thought the Earth was perfectly spherical. Hooke returned to the theme in the lectures he gave late in 1686 and early in 1687, to answer criticisms he had received. He demonstrated the processes taking place in two experiments. One involved simply blowing a bubble of glass using a glass-blower's pipe. When the pipe was held still while the molten glass set, the bubble formed a hollow sphere. But when the pipe was twirled around while the glass was setting, the resulting bubble bulged out at its 'equator'. The second demonstration involved a bowl of water on a turntable. When the turntable was spun, the water sank in the middle and rose up the sides of the bowl. Note that this happens when the bowl and water are both rotating, so the water is not moving relative to the bowl, unlike the case of tea stirred in a cup, which also forms a concave surface. Like all good scientific ideas, this one made predictions that were confirmed by other experiments. If the Earth bulged at the equator as Hooke suggested, gravity would feel weaker there, so a pendulum clock would run slower at (or near) the equator than at European latitudes, unless the pendulum was shortened. Edmond Halley, as we have seen, observed exactly this phenomenon on his visit to St Helena. It had previously been noted by the Frenchman Jean Richer, but his results were not published until 1679, which may explain Hooke giving priority to Halley when he said in one of those lectures 'that Phaenomena do answer to this Theory, has been verfify'd, first by Mr. *Hally* at St. *Helena*, and since by the *French* in *Cayen* . . . 'twas necessary to shorten the Pendulum to make it keep its due Time'. Newton, however, emphasised Richer's observations of 1672 in Book Three of the *Principia*, without mentioning that they were not widely known until 1679. Drake suggests that he did this to give the impression that his own views on the shape of the Earth were formed by learning of Richer's results around that time, while the correspondence shows that he did no such thing. 'Here,' she writes, 'is another example of history allowing Newton to usurp the credit that rightfully belongs to Hooke.'

Hooke's vision of geological processes is, as he puts it, 'almost' circular. Uplift is followed by erosion, with sediments being laid down to form new rock, followed by more cycles of uplift and erosion. He was what would later be called a gradualist, or uniformitarian – someone who recognises that the features of the Earth around us, especially great mountain ranges, are produced by gradual processes, the same processes we see at work today, but operating over very long intervals of time. When Charles Darwin realised, as a result of an earthquake and uplift he experienced and observed in South America during his voyage on the *Beagle*, that this is the way mountains are made, it was the same realisation Hooke had come to more than 150 years before. But the 'almost' in Hooke's discussion of cyclic geological processes is crucially important. Hooke also realised that eventually everything wears out: the law of nature later enshrined as the second law of thermodynamics, described by the twentieth-century physicist Arthur Eddington as holding 'the supreme position among the laws of Nature'.*

'Cities, Countries, Shores, nay the Sea itself are the Slaves of Fate,' says Hooke. 'The very ground we stand on is it self unfixt . . each part changes and sinks into Ruine and Alteration'.

Late in his life, in a lecture delivered in July 1699, Hooke referred to the 'Subterraneous Flame or Fire, or Expression, call

---

* *The Nature of the Physical World*, Cambridge UP, 1928.

it by any name you please' that drove the geological cycles, and expressly said that in time the fuel for this flame would be 'consumed and converted to another Substance, not fit to produce any more the same Effect.' He foresaw what nineteenth-century thermodynamicists would call 'the heat death of the Universe'.

Hooke's ideas concerning the origin of the Earth and its early history evoked 'sliding, subsiding, sinking and changing of the Internal Parts of the Earth' as a result of which 'many submarine Regions must become dry Land, and many other Lands will be overflown by the Sea'. This offered an explanation not just of the location of marine fossils today, but possibly also of Noah's flood. This 'terraqueous globe' idea did not catch on when Hooke presented it in front of the Royal Society in the early 1660s, but surfaced in a curious way nearly thirty years later.

John Aubrey, who was a good friend of Hooke, was so taken with the idea that when he wrote his *Memoires of Natural Remarques in the County of Wilts* in the 1680s he included a chapter on 'An Hypothesis of the Terraqueous Globe', subtitled 'A Digression'. He identified Hooke as the originator of the idea, but noted that it had been 'by many perhaps forgotten', so that 'I doe here hand [it] downe to Posterity (if this Essay of mine lives) with a due acknowledgement of his great Discovery'. In 1691, Aubrey sent a copy of the manuscript to the naturalist John Ray, the person who established the concept of species, for his advice about publishing it. Ray replied that he liked what Aubrey had written, but thought that the 'Digression' was 'aliene from your subject, & so may very well be left out'. Aubrey clearly disagreed. He wrote on the letter 'This Hypothesis is Mr. Hooks. I say so, and it is the best thing in the Book'.* Aubrey's book did not appear in his lifetime – it was eventually published as *Natural History of Wiltshire* in 1874, without the chapter on the Terraqueous Globe, a decision made by the posthumous editor, not by Aubrey. But in 1692, less than a year after he had read Aubrey's manuscript,

---

* Quoted by Drake, from Robert Gunther, *Further Correspondence of John Ray*, Ray Society, London, 1928.

Ray rushed into print his own *Miscellaneous Discourses Concerning the Dissolution and Changes of the World*, which was with the printer in January 1692. It contained the text of a sermon he had preached thirty years earlier, an 'explanation' of fossils as stones formed by the Deluge, not the remains of living things, and a Digression (his term) on the development of the early Earth which followed the same lines as Hooke's Terraqueous Globe idea, and cited the same references. But it did not cite Hooke. Aubrey was incensed, and wrote to a friend, Anthony Wood:

> Your advice to me was prophetique, *viz*, not to lend my Mss. You remember Mr J. Ray sent me a very kind letter concerning my *Naturall History of Wilts*: only he misliked my Digression, which is Mr Hooke's Hypothesis of the terraqueous Globe whom I name with respect. Mr. Ray would have me (in the letter) leave it out. And now lately is come forth a book of his in 8° which all Mr. Hooke's hypothesis in my letter is published and without any mention of Mr. Hooke or my booke. Mr. Hooke is much troubled about it. 'Tis a right Presbyterian trick.*

With great restraint, Drake notes that 'the circumstances seem suspicious'. It was, she says, 'an age of free-for-all piracy of ideas', giving the lie to the widespread accusations of paranoia levelled at Hooke's complaints about plagiarism. Hooke had more (and better) ideas than other people, so it is hardly surprising that he was a victim of this piracy.

Ironically, in one case the suggestion of plagiarism seems inappropriate, because the ideas of the supposed plagiarist were decidedly inferior to Hooke's. He was Niels Stensen, a Danish physician usually referred to as Steno, who has a place in the history books as the founder of the science of geology, based on rather limited evidence and a widespread ignorance of the significance of Hooke's earlier ideas.

---

* Anthony Powell, *John Aubrey and His Friends*, Hogarth Press, London, 1988.

By the time Steno, who was born in 1638, qualified as a physician, Hooke had already published *Micrographia* and given the talks on the Terraqueous Globe idea that Aubrey mentions. In the mid-1660s, Steno identified so-called tonguestones as the petrified remains of sharks' teeth. Because these were found in rock layers far inland today, he argued that the rocks themselves had been laid down underwater at various times in a series of floods, the latest of which was the biblical Deluge. In 1669, Steno published his only significant scientific work, with a title that translates as 'Predecessor of a dissertation of a solid naturally contained within a solid'. The solids within solids were, of course, fossils. But the promised full-length dissertation never appeared, because Steno became a priest; not just an ordinary priest, but an ascetic who took self-deprivation to extremes and gave everything he had to the poor. He abandoned science completely, possibly because he realised that science could not be reconciled with the literal interpretation of the Bible.

Did he learn anything from Hooke? Maybe. In a letter read at the Royal on 27 April 1687, Hooke claimed that Oldenburg had admitted to Hooke that he had 'transmitted the substance' of Hooke's early geological Gresham lectures to Steno, 'Some time before Mr Steno had published his Booke.' If so, however, Steno did not make good use of the information, since Hooke went much further. Hooke's real complaint was that Oldenburg, by getting Steno's book translated into English and promoting it, had contributed to the elevation of Steno and the overshadowing of Hooke. He had a point.

But although Hooke's geological thinking was overshadowed in his lifetime, and for a little while afterwards, within fifty years of his death it was widespread knowledge, in no small measure thanks to the publication of the *Posthumous Works*. Among a great deal of evidence presented by Drake to show how widely Hooke's work on 'earthquakes' was known in the second half of the eighteenth century, one stands out. Rudolf Erich Raspe, who is remembered today as the author of *Travels of Baron Münchausen*, but who was a serious geologist, published a book in 1763 whose title is translated by Drake as 'A Model of the Natural History of the Terraqueous globe, especially about new Islands born out of the sea, and from their exact descriptions and observations, further corroborating, the Hookian hypothesis of the Earth, on the origin of mountains and petrified bodies'. On the strength of this work, Raspe was elected as a Fellow of the Royal Society.

The most significant influence of Hooke's writings on this topic was, however, on James Hutton, the 'father of modern geology'. Hutton wrote a book, *Theory of the Earth*, published in 1795, which contained a wealth of observational facts and a carefully argued case for, among other things, gradualism and the immense age of the Earth; but his writing style was so impenetrable that it might have made no impact if his acolyte John Playfair, had not presented a beautifully clear summary of Hutton's work in his own book, *Illustrations of the Huttonian Theory of the Earth*, published in 1802. It was Playfair's book that kick-started the nineteenth-century geological revolution, most notably through its influence on Charles Lyell.* But who were Hutton's influences?

Playfair tells us that in his search for geological information and other interesting natural phenomena Hutton had 'carefully perused almost every book of travels from which anything was to be learned concerning the natural history of the Earth'.† Such a voracious reader surely cannot have missed two travel books, written by J. J. Ferber and by I. von Born, that were popular at the time; they had been translated into English and published by the Royal Society. The translator was Rudolf Erich Raspe, FRS, and in each of them he took the opportunity to give an overview of Hooke's ideas, quite apart from pointing the interested reader towards Hooke's own writings. And surely Hutton must have seen Raspe's own book. Another widely read eighteenth-century

---

* Lyell himself did refer to Hooke directly. In his *Principles of Geology*, published in 1832, he described Hooke's *Earthquakes* as 'the most philosophical production of that age, in regard to the causes of former changes in the organic and inorganic kingdoms of nature.'

† 'Biographical Account of the Late James Hutton, F.R.S.', *Transactions of the Royal Society of Edinburgh*, volume V, pp 39–99.

book, published anonymously, was *The History and Philosophy of Earthquakes, from the Remotest to the Present Times: Collected from the best Writers on the Subject*. This appeared in 1757, and nearly a third of the volume (106 pages out of 334) deals with Hooke's work.

Hutton, it seems, must have been aware of Hooke's ideas about 'earthquakes'. The points of agreement between Hooke and Hutton include, as Drake points out, 'the cyclic nature of sedimentation and denudation', as well as the whole concept of gradualism. Hooke, though, had the better grasp of deep time. In a paper published in 1788, Hutton wrote that 'we find no vestige of a beginning – no prospect of an end.' His cycles are perfect. But Hooke's version was *almost* cyclical. He envisaged the younger Earth as more dynamic and active, and the future Earth being less active and in effect dying.

Drake also draws attention to points of disagreement between the two where Hutton is at pains to dismiss certain ideas, such as polar wandering, which he has no need to introduce into his theory at all. 'One might wonder with whom Hutton is arguing,' she says, and concludes that he is arguing with the (unnamed) Hooke.

So we can trace a direct line from Hooke to Hutton (perhaps via Raspe), to Lyell (via Playfair), and to the whole of modern Earth science. Charles Darwin used to say that Lyell had given him 'the gift of time' – a long geological timescale, sufficient for the processes of evolution by natural selection to do their work. He seems to have been unaware that the first person to begin to understand the extent of geological deep time was also the first person to begin to understand the evolution of life on Earth, although that person had no inkling of the process of natural selection.

It is almost overkill, but cannot be allowed to pass without mention, that Hooke was also a pioneer thinker when it came to the circulation of the atmosphere of the Earth. Indeed, he seems to have been the first person to develop a proper model of atmospheric circulation at all, although his ideas tied in with the observations made by Halley at St Helena. Hooke inferred from travellers' tales of mist and fog at high latitudes that 'in and near the Polar Regions' there is 'dense and heavy air'. By contrast, 'Hurricanes, Tornadoes' and the like suggested to him that in the Torrid Zone the atmosphere is 'much more extended upwards, and of the vaporous Parts carry'd to a much greater height than elsewhere.' He was more or less right, although not entirely for the right reasons, and concluded:

> From these Considerations also will follow a necessary motion or tendency of the lower Parts of the Air near the Earth, from the Polar Parts towards the Aequinoctial, and consequently of the higher Parts of the Air from the Aequinoctial Parts towards the Polar, and consequently a kind of Circulation of the Body of the Air, which I conceive to be the cause of many considerable *Phaenomena* of the Air, Winds and Waters.

This is another example of the kind of thing Hooke could produce almost as an aside.

As we have seen, daringly for the century he lived in, Hooke didn't just question the biblical timescale, in particular for the Deluge, but dismissed it.* A couple of hundred days could not have been long enough to lay down 'so many and so great and full grown Shells, as this which are so found'. For Hooke, when science could not be reconciled with the literal interpretation of the Bible, it was the literal interpretation of the Bible that had to give. In a deliberate paraphrase of the Royal Society's motto, he says that when 'Nature speaks or dictates' it is appropriate to '*Jurare in Verba*' and listen to what she says.† Referring to his observations of the cliffs on the Isle of Wight, he writes: 'the quantity and thickness of the Beds of Sand with which [fossils]

---

* In one lecture, he mentions without comment (notably without adverse comment) 'that the Chinese do make the World 88,640,000 Years old'.

† There is a maxim, '*jurare in verba magistri*', which translates as 'to swear by the words of the master'.

are many times found mixed, do argue that there must needs be a much longer time of the Seas Residence above the same, than so short a space [as 200 days] can afford'. He also reports that one of these beds, containing 'Oysters, Limpits, and several sorts of Periwinkle', extends along the cliff 'I conceive near half a Mile, and may be about sixty Foot or more above the high Water mark.'* He explains the conflict between fact and Bible 'history':

> The great transactions of the Alterations, Formations, or Dispositions of the Superficial Parts of the Earth into that Constitution and Shape which we now found them to have, preceded the Invention of Writing, and what was preserved till the times of that Invention were more dark and confused, that they seem to be altogether Romantick, Fabulous, and Fictious, and cannot be much relied on or heeded, and at best will only afford us Occasions of Conjecture.

This is exactly the modern view. It is clear that way back in prehistory there was a catastrophic flood, perhaps related to the melting of great ice sheets at the end of the latest Ice Age, that affected early civilisations. It is recorded in, for example, the story of Gilgamesh, as well as in the Bible. But what knowledge of this event has passed down to us does indeed seem 'Romantick, Fabulous, and Fictious'.

Hooke also drew attention to the presence in the fossil record of the remains of creatures that are not found on Earth today. Using modern terminology, he realised that species went extinct, and that new species took their place. He noticed, for example, the similarities between the modern nautiluses and the extinct ammonites, but was also well aware of their differences (such as the smooth shell of the nautilus and the corrugated shell of the ammonite). Ammonites had disappeared from the Earth, perhaps being replaced by nautiluses:

---

* This is exactly like the observations Darwin made in South America, which brought home to him the significance of geological uplift.

There have been, former times, certain Species of Animals in Nature, which in succeeding and in the present Age have been and are wholly lost; for neither have we in Authors any mention made of such Creatures, nor are there any such found at present, either near the places of their position (as on the Shores or Sea about this Island) nor in any other part of the World for ought we yet know.*

And:

There have been many other Species of Creatures in former Ages, of which we can find none at the present; and that 'tis not unlikely but that there may be divers new kinds now, which have not been from the beginning.

What would Darwin have made of these passages:

There may have been divers new varieties generated of the same Species, and that by the change of the Soil in which it was produced; for since we find that the alteration of the Climate, Soil and Nourishment doth often produce a very great alteration in those Bodies that suffer it.

And:

Certainly, there are many Species of Nature that we have never seen, and there may have been also many such Species in former Ages of the World that may not be in being at present, and many variations of those Species now, which may not have had a Being in former Times . . . it seems very absurd to conclude, that from the beginning things have continued in the same state that we now find them.

---

* As Drake notes, unlike some of his contemporaries Hooke was scrupulous about citing such 'Authors' and giving the sources of his information.

This is evolution at work, and not far off natural selection. Extinction and variation, along with natural selection itself, are key features of the Darwinian understanding of evolution. Darwin did have a copy of *Micrographia*, but as far as we know never saw Hooke's writings on earthquakes.

Hooke also appreciated that the geological record might be used to construct a chronology of prehistory. He made an analogy with the way coins found in ancient remains in different locations (such as Roman coins in the north of England) could reveal the influence of a particular ruler in that place at a certain time, suggesting that fossil remains found in different strata could reveal when the strata were formed:

> Tho' it must be granted, that it is very difficult to read them, and to raise a *Chronology* out of them, and to state the intervalls of the Times, wherein such Catastrophes and Mutations have happened; yet 'tis not impossible, but that, by the help of those joined to other means and assistances of Information, much may be done in that part of Information also.

Partly to that end, Hooke makes an eloquent plea for the establishment of a truly scientific collection of fossils, not just a cabinet of curiosities, 'where an Inquirer might peruse, and turn over, and spell, and read the Book of Nature'.

Much has indeed been done along those lines since Hooke wrote those words, starting with the work of William Smith, yet another candidate for the title 'father of English geology', whose work as a surveyor of canals made him familiar with rock strata and an expert at using fossils to tell which layers were older and which younger, although an absolute chronology in terms of years had to await the twentieth century and such 'other means and assistances of information' as radioactive dating. Smith's geological map of England was published in 1815.

Hooke's geological studies, together with his thoughts about the history of life on Earth, alone qualify him as one of the leading scientists of the seventeenth century; the fact that he was also a pioneering microscopist and telescope-maker, a leading architect, contributed at least half of the 'Newtonian' revolution in physics, and found time for a wealth of other activities, almost defies belief. Although Jardine has suggested that the shadow of Newton 'for ever frustrates the possibility of allowing Hooke real standing in the history of science', we prefer the summing up made by E. N. Andrade in a lecture at the Royal Society in 1949, and published the following year:*

All those who have gone direct to Hooke have conceived the highest admiration for his astonishing industry, his whole-hearted devotion to science, his inventiveness, his ingenuity, his fertility and his brilliant theoretical insight.

That has certainly been our experience.

CHAPTER TEN

TO COMMAND A
KING'S SHIP

Nobody knows where or when Edmond Halley gained the nautical experience to make him fit to command a king's ship, but gain it he certainly did. One of the earliest clues we have to Halley's career at sea actually concerns work beneath the waves, not above them. As early as 1689 (the same year he estimated the size of atoms of gold) Halley presented to the Royal a chart of the mouth of the River Thames, discussed plans for a lighthouse on the Goodwin Sands, and reported on the use of diving bells to enable people to work underwater. The production, seemingly out of the blue, of the Thames chart is intriguing. It was presented to the Royal in the summer, so presumably was a result of a survey carried out by Halley in 1688 or earlier in 1689. At that time, William III was the new king, and war with France was anticipated, so there were strict controls on shipping. But, of course, if war threatened, accurate charts of the Thames approaches were essential. Anything moving in the area of the Thames Estuary ought to have had official permission, but no mention is made in the records of any vessel that might have been used by Halley. Cook refers to this as the 'dog that did not bark', alluding to the Sherlock Holmes story, 'The Adventure of Silver Blaze'. 'Halley's activities from midsummer 1688 to midsummer 1689', Cook says, 'may have been secret then and remain mysterious now.'

Halley's diving work is less mysterious. Diving bells – airtight containers open at the bottom and big enough for a man to stand in – had already been used on various occasions, but the divers inside the bells still had to hold their breath when they swam out to work. Halley suggested, in a paper titled 'A Method of Walking under Water', a heavy diving bell on wheels that could be moved around the seabed while the men worked. By providing the diver 'with such boots as the fishermen use he may be cloathed and stand dry on the bottom of the sea', he said, but the device was never built. Halley had the opportunity to put some of his ideas into practice, however, in 1691, when one of the ships of the Royal African Company, the *Guynie*, sank off the coast near Pagham, in Sussex.

The *Guynie* carried a particularly valuable cargo, including gold and ivory. She was a fast and relatively well-armed ship, referred to as a frigate, although the terminology was not as precise as it later became. The Company (in the form of a governing body known as the Court of Assistants) arranged for Halley to see what could be done about salvaging the cargo, which is why he was in Pagham when the possibility of obtaining the chair of astronomy in Oxford came up. The arrangement seems to have been an informal one at first. Halley investigated the wreck using a diving bell, in which he descended himself, which was replenished with air from barrels brought down from the surface. The success of the technique led to the grant of a patent to Halley, two members of the Company, and a City financier for a technique for taking air down to a diving bell, but the Company had financial problems and it was not until February 1692 that the Court formally decided to investigate the task of salvaging the *Guynie*, and only in April 1693 that Halley and his colleagues were contracted 'for the recovery of what was sunk with the Guynie frigatt'. Which is why

Halley was still working on the project (off and on) almost up to the time he went to the Chester Mint; this may explain why the undated letter mentioned earlier in which Halley apologises for not being able to call on Newton to discuss that appointment begins 'I had [i.e. would have] waited on you on Saturday, but I was obliged to go on board my friggat', but the letter may alternatively refer to another project, the main subject of this chapter.

There is no evidence that any significant amount of treasure was ever recovered from the wreck, and some evidence of the difficulties involved in a note by Pepys in the Naval Minutes of 'Mr Halley's having his vessel taken from him by a privateer when he was at work in diving upon a wrack'. But Halley left a detailed account of his diving work with the Royal.

The diving bell, sometimes referred to as a 'tub', was a truncated cone, three feet wide at the top, which was closed, and five feet wide at the (open) bottom. A stopcock at the top could be opened to let stale air out, and there was a small window made of thick glass. The bell was sunk with weights totalling one and three-quarter tons, and had a bench running around the inside just above the bottom 'for the men below to sitt on when they should be cold and whereon a man might sett with all his clouths at any depth drie'. Halley's own description of a descent in the device echoes Hooke's experiments in a vacuum chamber:

> When we lett down this engine into the sea we all of us found at first a forceable and painful pressure on our Ears which grew worse and worse till something in the ear gave way to the Air to enter, which gave present ease, and at length we found that Oyle of Sweet Almonds in the Ears, facilitated much this admittance of the Air and took of the aforesaid pain almost wholly.

The pain was caused, of course, by the build-up of pressure as the bell descended and water was pushed up into it from below. Every fifteen feet or so, the lowering was halted while more air was added to push the water out by increasing the air pressure inside the bell. This was the technique involving barrels of air that Halley had invented. Each cask was of about forty gallons capacity, sealed by being covered in lead. If such a cask of air was simply opened inside the diving bell, with the air inside the cask at normal atmospheric pressure, high-pressure air from the bell would rush *in* to the cask. Halley's ingenious idea, which he called 'the principall invention I can bost of', was to have a stoppered hole at the bottom of the cask and a valve at the top. When the top of the barrel was inside the bell, the bottom was kept below the water level and the stopper was released, while the valve in the top was open. Water flowing into the barrel forced the air out into the bell, and when the cask was full of water the valve was closed and it was hauled back to the surface. This was repeated until the water level inside the bell had dropped to the desired level. The technique worked:

> By this means I have kept 3 men 1¾ [hours] under the water
> in ten fathoms* deep without any the least inconvenience
> and in as perfect freedom to act as if they had been above.

Halley also designed a diving suit, made of leather with a helmet attached by hoses to the diving bell to provide air. And he noted the changed appearance and colours of objects underwater, and the way sounds were transmitted. Newton mentioned the optical effects in his *Opticks*:

> . . . the upper part of [Mr Halley's] Hand on which the Sun
> shone directly through the Water and through a small Glass
> Window in the vessel appeared a red Colour, like that of a
> Damask Rose, and the water below and the under part of
> his Hand illuminated by light reflected from the Water below
> look'd green. For thence it may be gathered, that the Sea-
> Water reflects back the violet and blue-making Rays most
> easily and lets the red-making Rays pass most freely and
> copiously to great Depths.

---

* A fathom is about 1.8 metres.

It seems the scientific aspects of the diving were more successful than the salvage.

Even while Halley was involved in this inshore work in the English Channel, plans were being laid for a much more ambitious nautical venture. The first we hear of it is in a rather cryptic entry in Hooke's diary on 11 January 1693, which mentions talk of Halley 'going in Middleton's ship to discover [that is, explore].' The Middleton referred to was Benjamin Middleton, who had been elected as a Fellow of the Royal Society in 1687 and was the son of a Colonel Thomas Middleton, who had died in 1672 but had been a Commissioner of the Admiralty, a friend of Samuel Pepys, and the man in charge of the dockyards at Chatham and Portsmouth at different times in his career. It seems that the younger Middleton was to be the financial backer for the project, offering to pay for the 'Victualls and Wages' if the Admiralty provided a ship, while Halley was the experienced sailor who would undertake the voyage, perhaps with Middleton on board. There is more detail in the Royal's archive, where there is a petition for the Royal:

> please to Lend their Assistance and Good offices to Obtaine of their Matys a vessel which may be Secure in all weathers, but not exceeding 60 Tunns burthen for a voyage to be undertaken by Benjamin Middlleton Esqr and Edmond Halley in order to discover . . . And the said Benj: Middleton for promoting the said Undertaking does oblige himself to goe [pay for] the Voyage and to Victuall and Man the said Vessell . . And the Care of Making the Necessary Observations is undertaken by the sd Edmund Halley, whose Capacity for Such Purposes is Supposed to be Sufficiently knowne . . .

Hooke's diary entry for 12 April 1693 is a little more forthcoming than his earlier comment: 'Hally & Middleton made proposals of going into ye South seas & Round the World'. This was the day the Royal Society formally endorsed the proposal, and the official records pick up the story on 12 July 1693, when the Commissioners of the Admiralty wrote to the Navy Board that:

Her Maty* is graciously pleased to incourage the said under-takeing. And in pursuance of Her Mats pleasure Signified therein to this Board. We do hereby desire and direct you forthwith to cause a Vessel of about Eighty Tuns Burthen to be set up and built in their Mats Yard at Deptford as soon as may be, and that Mr Middleton be consulted with about the conveniences to be made in her for Men and Provisions, and that when she is built She be fitted out to Sea. and furnished with Boatswains and Carpenters stores for the intended Voyage, & delivered by Inventory to the said Mr Middleton to be returned by him when the Service proposed shall be over.

But what was the 'Service proposed' that drew such an enthusiastic response from the Admiralty? Nothing less than a survey of the variation of the Earth's magnetic field from place to place across the seas, with a view to improving navigation – clearly, Halley's brainchild.

The ship – the first vessel to be constructed specifically for the purpose of a scientific investigation – was ready to be launched in April 1694. It was one of a class known as a Pink, with a shallow draught for inshore work (originally a design from the Netherlands) and bulging sides, fifty-two-feet long, a beam of eighteen feet, and a draught of just nine feet seven inches;[†] the displacement was roughly eighty-nine 'Tuns Burthen', not eighty, and she was named *Paramore*.[‡] The ship, it seems, was launched and ready to go – the other possible reason for the reference in Halley's undated letter to him needing to go on board 'my frigatt'. But soon after this the trail goes cold, and nothing more is heard of the project for two years. Perhaps the war with France made

---

* The King was abroad at the time.

† For comparison, the *Beagle*, on which Charles Darwin sailed, was 242 tons, just over ninety feet long, twenty-four feet six inches in the beam, and drew twelve feet six inches. *Beagle* was small; *Paramore* was tiny. The dimensions given here are from the Deptford Yard records, dated 14 May 1694.

‡ Official documents give the name as *Paramour*, but we follow the spelling Halley used in his writings.

it too risky a venture, especially in the light of Halley's encounter with a privateer, which may be why Halley took the job at Chester to fill in the time. It seems less likely that the job in Chester was attractive enough to be the reason for him postponing the voyage.

Whatever the reasons, the last mention of the *Paramore* in 1696 comes in August that year. Halley had been formally commissioned as Master and Commander of the vessel on 4 June, arrangements for funding the voyage had been made, and on 19 June Halley had drawn up a list of the proposed crew of the ship which 'with myself, Mr Middleton and his servant will be in all twenty persons', but then, on 15 August, out of the blue, the Admiralty ordered the ship to be laid up in wet dock awaiting further orders:

> We do hereby desire and direct you to cause his Majestys Pinke the Paramour to be laid up in the wett dock at Deptford until further order, notwithstanding any former directions to the contrary.

Nothing more is heard of Middleton at all, and nothing more of the *Paramore* until 1698 and her loan to Czar Peter. But the story from then on is comprehensively known, from Halley's own journals and the many official letters, reprinted in Norman Thrower's account of the three voyages of Edmond Halley in the *Paramore*.[*]

With the Nine Years' War over, and Peter having left, by the summer of 1698 plans were at last going forward for Halley to take the *Paramore* to sea to 'discover'. The ship turned out to need some modifications to improve her sailing qualities, but on 9 August the Admiralty ordered that:

> Whereas his Majesty has been pleased to lend his Pink the Paramour to Mr. Hawley for a Voyage to the East Indies or South seas, Wee do hereby desire and direct you, to cause

---

[*] The original log is in the British Museum.

her forthwith Sheathed and Fitted for such a Voyage, and that shee be furnished with Twelve Monthes Stores proper for her.

The ambitious idea of a round-the-world voyage, seven decades before the voyages of Captain Cook, seems, probably wisely, to have been dropped. But this is clearly now Halley's solo project, and on 19 August his Commission as Master and Commander was renewed, with his name entered in the wages book of the navy.

This appointment was remarkable in itself. Halley was not travelling as a passenger on a ship commanded by a naval officer. He had been appointed as a naval officer, to take full command of the ship. This is the only recorded occasion that such a 'landsman' was given such a command. The crew was being supplied by the navy, but Halley was in command of the expedition and would be giving orders to the crew, so he had to be a naval officer entitled to give them orders. He became known as Captain Halley, because the commander of a Royal Navy ship is always given the courtesy title Captain, whatever his substantive rank.*

Over the next few weeks the crew was completed while the stores and the modest armament of the *Paramore* (six three-pounders and two swivel guns) were installed. There was one other naval officer, Edward Harrison,† a Lieutenant who was officially second in command, but seems to have been expecting to be allowed a free hand to run the ship, and a Midshipman, John Dunbar. The Lieutenant, in accordance with usual practice, would be the navigating officer of the ship, responsible for making sure it got to the places the Captain ordered. The Boatswain/ Gunner, John Dodson, was the most senior of the other ranks, and there was a surgeon, George Alfrey. Halley himself drew up the official orders for the expedition, which the Admiralty then issued to him formally on 15 October 1698, addressed to 'Captn.

---

* 'Captain' James Cook, for example, was actually a Lieutenant when he made his famous voyage.
† No relation to the clockmaker.

Edmd Halley Commandr of his Mats Pink the Paramour'. They clearly spell out the scientific purpose of the voyage:

You are to make the best of your way to the Southward of the Equator, and there to observe on the East Coast of South America, and the West Coast of Affrica, the variations of the Compasse, with all the accuracy you can, as also the true Scituation both in Longitude and Latitude of the Ports where you arrive. You are likewise to make the like observations at as many of the Islands in the Seas between the aforesaid Coasts as you can (without too much deviation) bring into your course: and if the Season of the Yeare permit, you are to stand soe farr into the South, till you discover the Coast of the Terra Incognita, supposed to lye between Magelan's Streights and the Cape of Good Hope, which Coast you are carefully to lay downe in its true position. In your return home you are to visit the English West India Plantations, or as many of them as conveniently you may, and in them to make such observations as may contribute to lay them downe truely in their Geographicall Scituation And in all the Course of your Voyage, you must be careful to omit no opportunity of Noteing the variation of the Compasse, of which you are to keep a Register in your Journal.

Phew! All on a ship fifty-two-feet long with a crew of twenty, including Halley.

The voyage began on 20 October, but quickly ran into teething troubles. Rough weather in the Channel revealed gaps in the badly caulked seams, so the ship took in water, and then the sand used as ballast clogged the pumps when the water was being pumped out. Halley had to put in temporarily to Weymouth (where, typically, he took astronomical and magnetic observations), then went back to Portsmouth to have the gaps in the planking recaulked and the sand ballast replaced by shingle. Bad weather then delayed the expedition, but on 29 November *Paramore* sailed again, this time in company with a squadron commanded by Rear

Admiral John Benbow, on his way to the West Indies, which gave them protection from any possible pirate attack until they parted company with Benbow at Madeira. It was after they were left to their own devices that Halley began to have trouble – not with pirates, or with the ship, but with his crew.

Carrying out his orders to the letter, by early February 1699 Halley was heading for the island of Trinidada, in the Atlantic at 20° South, when the ship became becalmed in the Doldrums for so long that water ran low. So he altered course for Fernando Loronho, at 4° South. But on 17 February, while the Boatswain was on watch, between two and three in the morning Halley looked out to find that the Boatswain was steering north-west, instead of heading due west for the island. He concluded that this was a deliberate 'designe to miss the Island, and frustrate my Voyage'. With the course corrected, they sighted the island some twelve hours later, and made landfall the following day. After a brief stop, they reached Brazil on 26 February, the first British ship to visit the region of the Paraiba River, a few degrees south of the equator, in thirty years. Halley took advantage of the stop to make a chart of the region, and correct errors in the existing maps, something he did at almost every stop on his voyages. With the southern hemisphere winter approaching, Halley decided not to venture further south, and after replenishing the ship headed northward for Barbados, hoping among other things to exchange the more troublesome crew members (the hope was not fulfilled). It was on this leg of the voyage that matters came to a head.

With Barbados in sight, Lieutenant Harrison, who had the watch, became not only disobedient but insolent:

pretending that we ought to go to Windward of the Island, and about the North end of it, whereas the Road is at the most Southerly part almost. he persisted in this Course, which was Contrary to my orders given overnight, and to all sense and reason, till I came upon Deck; when he was so farr from excusing it, that he pretended to justifie it; not without reflecting Language; about 6 I commanded to bear away NW and NWbN and before 11 we came to an Anchor in Carlisle bay.

Although the unpleasantness did not stop Halley making his observations of longitude, magnetic variation, and tides, at various islands, he eventually relieved Harrison of his duties and took over the navigation of the ship himself. If Harrison expected Halley to make a mess of the job, he was sadly mistaken. Hardly surprisingly, the expert astronomer proved well equal to the task, and after leaving the West Indies on 9 May 1699 had no difficulty in making landfall at the Scilly Isles on 20 June and reaching Plymouth on the 23rd.

The same day Halley wrote to the Admiralty to explain his early return, describe his successes, and request an opportunity to carry out another voyage, leaving earlier in the season in order to venture further south. This letter gives us more information about Harrison's behaviour:

I this day arrived here with his Maties Pink. the Paramore in 6 weeks from the West Indies, having buried no man during the whole Voiage, and the Shipp being in very good condition. I doubt not but that their Lopps will be surprised at my so speedy return, but I hope my reasons for it will be to their satisfaction. For as, this time, it was too late in the year for me to go far to the Southwards . . in case their Lopps, as I humbly hope, do please that I proceed again for I find it will be absolutely necessary for me to be clear of the Channell by the end of August or at farthest by the middle of September. But a further motive to hasten my return was the unreasonable carriage of my Mate and Lieutenant, who, because perhaps I have not the whole Sea Directory so perfect as he, has for a long time made it his business to represent me, to the whole Shipps company, as a person wholly unqualified for the command their Lopps have given me, and declaring that he was sent on board here because their Lopps knew my insufficiency . . . he was pleased so grossly to affront me, as to tell

me before my Officers and Seamen on Deck, and afterwards
owned it under his hand, that I was not only incapable to
take charge of the Pink, but even of a Longboat; upon which
I desired him to keep his Cabbin for that night, and for the
future I would take charge of the Shipp myself, to shew him
his mistake: and accordingjy I have watcht in his steed ever
since, and brought the Shipp well home . . Notwithstanding
that I have been defeated in my main design of discovery, yet
I have found out such circumstances in relation to true
Variation of the Compass, and the method of observing the
Longitude at Sea (which I have severall times practised on
board with good success) that I hope to present their Lopps
with something on those articles worthy of their patronage.

Inevitably, Harrison was tried by a Court Martial, on 3 July,
with Sir Cloudsley Shovell as President of the Court. Perhaps
equally inevitably, given that he was a professional seaman being
tried by professional seamen who could not regard Halley as one
of their own, he got off with a 'Severe reprimand'. This seems
like a light punishment, but Ronan suggests that it was all that
the Court could impose, because the only charge actually brought
against Harrison was insolence.

In a letter to Josiah Burchett, Pepys' successor as Secretary to
the Admiralty, Halley grumbled that:

I fully proved all that I had complained of against my
Lieutenant and Officers, but the Court insisting upon
my proof of actuall disobedience to command, which I had
not charged them with, but only with abusive language and
disrespect, they were pleased only to reprimand them, and in
their report have very tenderly styled the abuses I suffered
from them, to have been only some grumblings such as
usually happen on board small Shipps.

In any case, Harrison soon resigned his commission and joined
the merchant service, while Halley retained his naval rank.

In the summer of 1699, Halley attended meetings of the Royal (where his successor as Clerk, Israel Jones, had been appointed on 8 March 1699), and among other things presented the Society with a chart plotting his magnetic observations, and some corrections to the map of Brazil. But his wish for a second voyage was soon granted, and although the *Paramore* was paid off on 20 July, on 24 August the first names of a new crew were entered in the wages book. This time, the crew would number twenty-four, including a few of the more reliable hands from the first voyage, with Halley as the only officer and in complete command – as, indeed, he was on the first voyage, but now with nobody to dispute his authority. The ship sailed on 16 September, and proved to handle much better, having been refitted during the summer. Halley's orders, essentially the same as for the first voyage, now included the command to 'proceed to make a Discovery of ye unknowne southlands between ye Magellan Streights and ye Cape of good hope between ye Lattd of 50 & 55 South', without, of course, neglecting the scientific observations.

This time, *Paramore* was escorted as far as Madeira by an armed merchantman of the Royal African Company, the *Falconbird*, and as well as his other observations Halley was now keeping records of temperature and atmospheric pressure. In rough seas off Madeira (too rough for the ship to put in for wine), a boy named Manly White, entered on the books as one of the Captain's servants, was lost overboard. But there were no further tragedies as the ship pressed on southward, leaving the track of the first voyage, crossing the equator at Longitude 23° West (of London) on 16 November, and reaching Rio de Janeiro, at 22° 55' South, on 14 December.

After replenishing his stores, in accordance with his orders Halley left Rio for the deep south on 29 December 1699, right at the start of the southern summer. The voyage south might be described as uneventful, as far as a naval captain was concerned, although mixed in with mentions of 'very fair weather' the journal contains entries such as this one for 14 January 1700:

Yesterday in the afternoon it blew a hard Gale at N N E and towards Evening it Came to North and blew so hard that we were obliged to Scudd before it: about Nine it began to lighten [lightning] and from ten till half an hour past eleven we had Terrible Thunder lightning and rain with vehement Squalls of wind . . . By twelve the Storme was over and it began to clear up and a fine Gale sprung up at W S W. so I ordered the Sailes to be sett and to goe away S S E with the Wind on the beame.

And this was still at a latitude of 37° South. Fans of the Hornblower or Aubrey novels will be able to read between the lines and appreciate the hazardous nature of this voyage to high southern latitudes in a small ship, and the seamanship required to carry out Halley's orders.

By 27 January the ship had passed 50° South, and Halley records seeing two kinds of penguins, making him think that they were near land. The next day, he saw 'a Couple of Annimalls which some supposed to be Seals but are not soe; they bent their Tayles into a sort of a Bow'. The description, and a little drawing he inserted of the shape of the creatures' tails, clearly indicates that they were whales. By now they were beginning to experience extreme cold, with the air temperature below the freezing point of water even though it was high summer. On 1 February, at 52° 24', their furthest south and some 35° West of London, the ship encountered what looked at first to be three large islands, 'all flatt on the Top, and covered with Snow. milk white, with perpendicular Cliffs all round them.'* Cautiously closing in to investigate, Halley was hampered both by strong winds and by fog. The ship was now in real danger, as Halley records with a certain amount of sang-froid on 2 February:

between 11 and 12 this day we were in imminent danger of loosing our Shipp among the Ice, for the fogg was all the

---

* They were then almost due north of the region of Antarctica now known as Halley Bay.

morning so thick, that we could not See a furlong about us, when on a Sudden a Mountain of Ice began to appear out of the Fogg, about 3 points on our Lee bow: this we made a Shift to weather when another appeared more on head with several pieces of loose Ice round about it; this obliged us to Tack, and had we mist Stayes, we had most Certainly been a Shore on it.

There were further alarms of this kind before the ship got free from the ice and Halley, his orders carried out, headed back north.

This danger made my men reflect on the hazards wee run, in being alone without a Consort, and of the inevitable loss of us all, in case we Staved our Shipp.

It wasn't just the major icebergs that were the problem, but lesser chunks of floating ice which surrounded the ship and which could indeed have staved in the sides of the fifty-two-foot-long wooden ship.

Heading north again, Halley had time to reflect on what he had seen. The icy masses he had encountered were not themselves land, but he knew, as Hooke had explained to the Royal during the severe winter of 1683–84, that ice floats with only a tiny part of itself above water, and could not imagine that what he had seen represented no more than an eighth of a great mass of floating ice. He concluded that they must be masses of ice grounded on land beneath the waves, and when he was able to report the discovery in a letter to the Admiralty at the end of March he wrote:

we fell in with great Islands of Ice, of soe Incredible a hight and Magnitude, that I scarce dare write my thoughts of it, at first we took it for land with chalky cliffs, and the topp all covered with snow, but we soon found our mistake by standing in with it, and that it was nothing but Ice, though it could not be less than 200 foot high, and one Island at least 5 mile in front, we could not get ground in 140 fathom. Yet I conceive it was aground, Ice being very little lighter than water and not above an Eight part above the Surface when it swims; It was then the hight of Summer, but we had noe other signe of it but long Days; it froze both night and day, whence it may be understood how these bodies of Ice are generated being always increased and never thawing.

Halley's conclusion was incorrect, and it is clear to us that what he saw must have been huge masses of floating ice that had broken free from the ice shelf around Antarctica (some of which is, indeed, grounded on land below the waves). But, as ever, he drew an entirely reasonable conclusion based on a scientific assessment of the facts available to him.

The ship did not get free of the ice, having been further south in the open Atlantic than anyone before them, until 5 February.* They then headed northward in gale-force winds 'to recover the warm Sunn'. Halley intended to make for the Cape of Good Hope, passing by Tristan da Cunha without stopping, but as the ship was driven northwards by the wind this proved impossible, and he changed course for St Helena to reprovision. The dangerous part of the voyage was far from over. On 26 February Halley recorded 'daylight with a Terrible high Sea about Six this Morning a greate Sea broke in upon our Starboard quarter, and withall threw us to that we had likt to have oversett; the Deck being full of Water, which had a clear passage over the Gunnell, but it pleased God She wrighted again'. They arrived at St Helena, battered and very low on both food and water, on 12 March, two-and-a-half months after setting sail from Rio, and just over twenty-three years since Halley first set foot on the island. They left again on 30 March, heading for the island of Trinidada, which Halley had been unable to visit on his first voyage. Trinidada, which they reached on 15 April, lies about 750 miles east of the Brazilian coast

---

* If they had not had to turn back when they did, they might have discovered the islands now known as South Georgia, a little further south, which were found by James Cook on his second voyage.

at 20°31'30''S 29°19'30'' W. It turned out to be an uninhabited lump of rock (actually the largest lump of rock in a small archipelago), but with one asset: a good supply of drinkable water, which made it a desirable port of call for mariners. Halley formally claimed the island for the Crown, and left a breeding stock of hogs and goats to provide food for any future visitors. The descendants of these animals survived into the twentieth century, unlike the British claim to the island, which now belongs to Brazil. But Trinidada offered no facilities for overhauling the ship, and on 20 April Halley set sail for Pernambuco (now Recife) on the Brazilian coast, going ashore there on the 29th.

It was here that one of the most bizarre events of the voyage occurred. A Mr Hardwick, who styled himself British Consul but actually held no such post and was merely a representative of the Royal African Company, initially refused to believe the evidence of Halley's two commissions and suspected the *Paramore* of being a pirate ship. On what he thought would be a social visit to Hardwick's house, Halley was detained under guard while his ship was searched and some of his sailors were interrogated. 'But Finding no Signes of Piracie on Board he came and discharged me of my Guard begging my pardon.' Hardwick claimed to have been acting on the instructions of the Portuguese Governor, although Halley had found that man to be 'very obligeing'. Hardly surprisingly, the ship left Pernambuco as soon as possible, on 4 May, bound for the northern hemisphere.

On 21 May Halley reached Barbados, and went ashore to visit the Governor, but was advised to leave at once because disease was rife on the island. Although they hurried away, Halley himself and several crew members became ill. 'The Barbadoes disease,' Halley wrote, 'in a little time made me So weake I was forced to take [to] my Cabbin' while the Mate steered a course for St Kitts, referred to by Halley as St Christophers. So it was there that Halley, recovering from his sickness, was able at last to have his battered little ship partly overhauled, with new rigging and other repairs. A further stop at Anguilla gave an opportunity for replenishing the ship's stores for the forthcoming Atlantic crossing and for the crew to have some shore leave. Then on to Bermuda, reached on 20 June, to have the overhaul completed with the ship careened, cleaned, caulked and painted. With everything as well set up as possible, on 11 July the *Paramore* stood out to sea for the journey home, initially heading north-north-east for Cape Cod to take advantage of the prevailing currents and wind.

The weather prevented any chance of a landing on the coast of New England, but at the end of the month the ship had reached Newfoundland, where in thick fog they might have run aground 'had we not fell in with some French Fishermen' who put them on the right course. An English fishing fleet, however, fled at their approach, taking them for a pirate ship, and one of the boats took some pot shots at the *Paramore*, but with no harm done. This was the last alarm of the voyage. After taking on fresh water, on 7 August the ship weighed anchor and set course for England, passing Scilly lighthouse on the 26th, and on 10 September Halley 'Deliver'd the Pink this evening into the hands of Captn William Wright Mastr of Attendance at Deptford'. The boy Manly White was the only crew member lost on either voyage. In six days short of a full year, Halley had achieved all his objectives and fully earned Pepys' accolade as 'the most, if not to be the first Englishman (and possibly any other) that had so much, or (it might be said) any competent degree (meeting in them) of the science and practice (both) of navigation.'

On 30 October Halley showed the Fellows of the Royal Society a map of his magnetic observations, and entertained them with an account of the islands of ice that he had seen. Halley's charts were the first to show isogones – lines joining points of equal magnetic deviation* – and his mapping of the coasts around the Atlantic was better than anything produced previously. The magnetic data used for the charts included not only Halley's observations, but also information he had been gathering for years from other sources. He was not only a first-rate navigator and explorer, but a geographical innovator. On 30 November, no

---

* Initially, they were known as 'Halleyan lines'.

longer Clerk, Halley was re-elected as a Fellow of the Royal Society. But his career as a naval captain was not yet over.

Halley had a long-standing interest in tides, and as long ago as 1684 had published a paper discussing the strange tides in the Bay of Tonkin. A resident of the region, Francis Davenport, had reported a peculiar pattern in the tidal flow at Batsha, on the south coast of China, in which, among other oddities, there was one high tide a day, except that every fortnight there was no tide at all. Halley realised that there was a connection with the Moon's position in its orbit, which is not a simple circle around the Earth, but in the *Philosophical Transactions,* where Davenport's observations were reported, said that 'to attempt to assign a reason, why the *Moon* should in so particular a manner influence the *waters* in this one place, is a task too hard for my undertaking'. Tides were also important to mariners, and Halley had been hampered on both his voyages by the inconvenience of dealing with the poorly understood tidal flows up and down the English Channel. So it is no great surprise that on 23 April 1701 he wrote to the Admiralty proposing

> That if their Lopps shall think fitting to have an exact account of the Course of the Tides on and about the Coast of England . . . there be provided a small Vessell such as their Lopps shall think proper . . . for which service their Lopps most obedient servant humbly offers himself.

Their Lordships did indeed think it fitting to have such a survey (of the Channel, which is all Halley intended in spite of the grand reference to the Coast of England), and naturally chose the *Paramore* for the task. Halley received his Commission on 26 April, so soon after he made his formal proposal that it seems there had already been discussions about the project. Bureaucracy and some difficulty in assembling a new crew delayed the task, but while this was going on, on 6 May Halley's previous achievements were acknowledged when the Admiralty made an order to the Navy Board that:

Wee do hereby desire and direct you to cause to be paid unto Captn Edward [sic] Halley, out of the Money in the hands of the treas of the Navy, upon Acct of the Tenths of Prizes the Sum of two Hundred Pounds in consideration of his great Paines and care in the late Voyage he made for the discovering of the Variation of the Needle.

The instructions for the new task (written, of course, by Halley himself) told him to:

use your Utmost care and Diligence in observing the Course of the Tydes accordingly as well in the Midsea as on both Shores As also the Precise times of High and Low Water of the Sett and Strength of the Flood and Ebb, and how many feet it flows, in as many, and at such certain, places as may suffice to describe the whole. And whereas in many places in the Channell there are Irregular and half Tydes you are in a particular Manner to be very careful in observing them.

   And you are also to take the true bearings of the Principal head Lands of the English Coast one from another and to continue the Meridian as often as conveniently may from Side to Side of the Channell, in order to lay down both Coasts truly against one another.

Work on the tidal survey began on 14 June, and kept Halley busy until 2 October. It was a tedious and painstaking task, carried out in small boats, often in poor weather, which there is no need to go into in detail here. But the success of the survey demonstrates Halley's skills as an organiser and a leader, a man who got things done. He produced the first detailed survey of the complex tides in the Channel, a full century before anything comparable was done by others, having already produced the first detailed study of the Earth's magnetic field. At the end of it all, in spite of the high esteem in which he was held at the Admiralty, Halley did not immediately receive his full wages, and there may have been some behind-the-scenes wrangling about his official status, because an Admiralty letter to the Navy Board, dated 20 February 1702, reads:

> Captain Edmd Halley late Comander of the Paramour Pink, having acquainted me that hee has not yet received his pay for his two last voyages; and for as much as he is not esteem'd* as a Captain in the Navy, but only Employd by Particular Order from His Majesty for the Improvement of Navigation; I do therefore hereby desire and direct you to cause him to be forthwith pay'd the Wages due to him for the time he Comanded the said Vessell.

And on 20 April 1702:

> It being her† Mats Royl Will and pleasure that the Summ of two hundred pounds shall be payd to Capt Edmd Halley (over and above his Pay as Captain of the Pink the Paramour) as a reward to him for his Extraordinary pains and care he lately took, in observing and setting down the Ebbing, and Flowing, and setting of the Tydes in the Channell as also and bearing of the head-Lands on the Coasts of England and France. I do therefore in obedience to her Maty Commands hereby desire and direct you, to cause the sayd summ of two hundred pounds to be payd unto Him the sayd Capt Halley accordingly.

The reference to the headlands of France is significant. Halley's log does not record details of a survey of the French coast, but it is clear from other references that he made one, which must have been presented to the Admiralty and which would be invaluable in the anticipated war with France.

This was the end of Halley's official connection with the Navy; although he talked of voyaging to the Pacific to complete the

---

* That is, not officially on the books; he was clearly esteemed as we use the term.

† Queen Anne had recently (on 8 March) acceded to the throne on the death of William III.

magnetic survey, nothing ever came of the plan. It was also the end of the *Paramore*'s naval service – she was sold by the Admiralty in 1706, for £122, having only ever had two Captains, Peter the Great and Edmond Halley. But it wasn't the end of Halley's activities for 'King (or Queen) and Country'.

Halley had probably been discreetly studying not just 'bearing of the head-Lands on the Coasts of England and France', but the approaches to French harbours, useful information in time of war. War with France was always expected in those days, and the political pot had been stirred by the death of Charles II of Spain in 1700 and the succession of Philip V, the seventeen-year-old grandson of Louis XIV of France, to the Spanish throne. This concentration of Bourbon power was too much for many other nations to tolerate, and triggered the War of the Spanish Succession, in which England, the Dutch and the so-called Holy Roman Empire (essentially an Austrian Empire) were allies. This was the war in which the Duke of Marlborough, arguably England's greatest-ever general, made his name, notably at the battle of Blenheim. But it also involved war at sea.

The English, Dutch and Austrians formally declared war in May 1702. Between them, the English and Dutch had powerful naval forces, but lacked a base in the Mediterranean and a way to link up by water with their Austrian allies at the head of the Adriatic Sea. George Stepney, the English Ambassador in Vienna, suggested that if an English, or Anglo-Dutch, squadron could be based in the Adriatic all year round it would be invaluable, and recommended that a suitable person be sent to find an appropriate port. The job description could have been written for Halley, who was the ideal man, both for his nautical experience and because he was a suitable Gentleman to be a representative of the Queen. His nautical rank of Captain was equivalent to Colonel in the army, which ensured that he would be well received by the protocol-conscious Imperial Court. And he was known to be discreet.*

---

* The most thorough account of Halley's work in the Adriatic has been provided by

Halley received his formal instructions in a letter from Queen Anne dated 4 November 1702; like his naval orders, the letter was clearly drafted by Halley himself:

You shall immediately upon receipt hereof repair to the Emperors Court at Vienna, or such other Place where he may be. and there apply your Selfe to Our tr. and Wellb. Stepney Esqr, Our Envoy Extrary or in his absence to his Secretary, whom We have directed to be assisting to You in obtaining such Commission or Orders [from the Emperor] as shall be sufficient to enable you to perform the Service, whereupon you are sent, and herein . . .

. . . you shall make what haste you can to the said severall Ports, and take exact Plans of each of them, making such Observations and Remarks as you think may any way conduce to Our Service.

You shall particularly sound the depth of water in each Port, what Rocks or Sands are in each Port, and in the Entrance into it, what number of Ships can ride there securely against Wind and Weather, what fortifications can be made for the defence of them against an enemy, and in what manner and with wheat charge the same can be made.

What conveniencys there are or may be made there for careening, cleaning and repairing Our Ships . . .

You are to go from hence to the Hague, and there communicate your instructions to the Earle of Marlborough . . .

You are to use all diligence in your journey to Vienna and thence to the Adriatick and returne and in the execution of these Our Instructions, since it is of great importance to Our Service, that we have as speedy an Account of these matters as is possible . . .

Halley, travelling with his servants, was in the Hague on 12 December, and moved on via Leipzig and Prague to Vienna, arriving on 10 January 1703. He could not travel more directly, through Bavaria, which was a French ally. In Vienna, Halley had to wait while the diplomatic niceties were observed until he could receive the formal documents allowing him passage to the Adriatic, and used his time to study the available maps and other information about the ports. He found only one that was likely to meet the requirements of a squadron of Her Majesty's ships – Buccari, a little to the south of Fiume. From Vienna, Halley moved on to Trieste, arriving on 1 February, and carried out a brief survey of the port, reporting back to Stepney that it was not suitable for their needs. But Buccari, which Halley reached on 10 February, offered everything he had hoped for, with a narrow and easily defended entrance opening up to a sheltered bay with plenty of anchorage. The entrance is only about 400 metres wide, but the bay itself is about a kilometre wide and four-and-a-half kilometres long, running parallel to the coast. After surveying the harbour and taking depth soundings, identifying the best sites for gun batteries to defend the harbour, and taking a quick look round the nearby region, Halley was back in Vienna on 27 February, where he used his observations to draw up charts of the harbour, one set of which was presented to the Emperor on 10 March. In return, Emperor Leopold gave Halley a diamond ring worth £60 (taken, the story goes, from the Emperor's own finger); a sensational sign of favour which is mentioned in all the accounts of Halley's visit to Vienna.* Just a week later, armed with his charts and data, he was on his way back to London (unfortunately, no copies of the charts survive).

Halley arrived home by the middle of April, carrying also a letter from Stepney reporting to the Secretary of State, the Earl of Nottingham, that:

> he has performed what he was sent about entirely to the satisfaction of this Court and I hope that your Lordp will herein be satisfied with the discoveries he has made.

---

* It was while Halley was occupied with this work that, on 2 March, his friend Robert Hooke died. Samuel Pepys died in the same year.

He had indeed carried out his duties with 'all diligence'; but the same could not be said of the Austrians left to carry out the work at Buccari. In pite of Stepney's prodding, very little work was carried out, and Halley was sent back to see what could be done to speed things up, leaving London on 22 June 1703 and arriving in Vienna on 23 July. On the way he passed through Hanover, and is reported to have met the Elector, who was Queen Anne's designated successor, and already known as George I; the future King clearly had an interest in the activities Halley was now involved in.

In Vienna, Stepney was relieved to see him:

> It will be an ease to me to have Capt. Halley in these parts to look after the Fortifications and Provisions which are things I do not understand and I should have been unwilling to rely on the assurances these Ministers can give me, that all is in readiness, wherein they are apt to deceive both themselves and others.

But even Halley was frustrated by the task in hand. The work, especially the key task of preparing batteries for the guns to defend the harbour, was proceeding slowly, both because those in charge believed (or pretended to believe) that no fleet would be coming that year, and because of a problem that has a familiar resonance today: the workers were, as Halley wrote to Nottingham, 'being well paid by the day while the work is in hand', so it was in their interest to carry it out slowly. When Halley did manage to get things moving on the construction of the batteries, another problem arose. Where were the cannon to go in them? The Austrians wanted the English to take some out of the ships and mount them in the batteries when the fleet arrived; the English wanted the Austrians to supply the guns. While the wrangling continued, a squadron under Sir Cloudsley Shovell did enter the Mediterranean in 1703, but stayed only briefly before returning home. Halley also returned to England in the autumn of 1703, having achieved less than he had hoped but probably more than anyone else could have done. The following year, the capture of Gibraltar by the English and the destruction of the French fleet at Toulon removed the immediate need for an Adriatic base, and the project was never completed.*

There is, however, an intriguing twist in the tail. Halley had arrived home in November 1703. A letter from the Earl of Nottingham dated 14 January 1704 instructs the Exchequer to pay to Stepney the sum of £36 'out of the secret fund' as a reimbursement for money he had given to Halley for expenses he had incurred.† Expenses doing what, we shall never know, but it seems that Halley's travels around Europe also provided the cover for a little quiet spying.

Whatever he had been up to, as far as his career was concerned, Halley's return to England had been perfectly timed. All thoughts of future voyaging seem to have disappeared with the news that John Wallis, the Savilian Professor of Geometry at Oxford, had died on 28 October 1703 (the day before Halley's forty-seventh birthday). It opened up the possibility of settling down into the academic life Halley had long wanted.

---

* It was as a result of this war that, among other things, Britain gained not only Gibraltar but also Minorca, and the whole of both Newfoundland and St Kitts, previously shared with France, while Spain lost the Netherlands to the Austrians.
† See MacPike.

CHAPTER ELEVEN

LEGACIES

Patronage was of key importance in deciding appointments to Oxford or Cambridge professorships, and at least one prospective candidate for the Savilian Chair moved quickly to call in a favour. On 6 November 1703 Viscount Hatton, of Kirby Hall in Northampton, wrote to the Earl of Nottingham, who was Secretary of State but a more junior member of the family which Hatton headed, mentioning that:

> I am so much obliged to Dr Keith . . . for his care of me in my late sickness that I cannot but recommend his brother, at his request, for the 'mathematique' Professorship at Oxford, for which he is a candidate. He is, I know, very capable of it, 'but if Mr Halley be thought of, or aim at it, he acquiesces and would by no means oppose it'.

Nottingham was obliged to let Dr Keith and his brother down gently, writing to Hatton on 20 November, with Halley not yet back in London, that:

I can't say yt Mr Halley thinks of ye Mathematick professor's place but everyone who has a vote in the election thinks of him; and [I] am very glad yr Lops recomendation of another is accompanyd with this condition, if Mr Halley does not pretend to it, for I have seen his zeal in the publick service (on which he is now abroad) the which added to his extraordinary skill above all his competitors obligd me to be very forward in promoting him to this place.

One person was not happy at the prospect of Halley's appointment. In a letter to Abraham Sharp, dated 18 December 1703,* Flamsteed grumbled 'Dr Wallis is dead – Mr Halley expects his place – who now talks, swears and drinks brandy like a sea captain'. Halley was indeed a sea captain, and proud of it. After his election, on 8 January 1704, he still preferred to be known as 'Captain' Halley, rather than as 'Professor' Halley. He was formally appointed to the post of Savilian Professor of Geometry on 7 March, and gave his inaugural lecture on 24 May 1704. He was now forty-seven, and had been elected to the Council of the Royal Society on his return to England, the same day that Newton was elected as President.

Halley's income as Professor was notionally about £150 a year, from rents on estates endowed by Sir Henry Savile back in 1619, but this seems to have been no more reliable than Hooke's income from Cutler and the Royal, and by the time Halley became Astronomer Royal he was owed more than £1,000. But his duties were not onerous: he had to lecture once a week in term time and to be available for one hour each week to give tuition to anyone who might call on him at the house in New College Lane which went with the job. Like other professors, however, he was often away from Oxford, at his home in London.

As with Hooke, Halley worked on different projects at the same time, and we have to unpick the threads and tell each story in turn. One important theme of his day, but of less interest now,

---

* Now in the University Library in Cambridge.

concerned his translation of Ancient Greek mathematical treatises, using Arabic copies of the original Greek texts. But it was Ancient Greek observations, not mathematics, that led Halley to a profound discovery, linked to an ongoing dispute between Flamsteed and the entire academic establishment of his day. Amazingly to modern eyes (and to most of his contemporaries), by the time Halley became a Professor in Oxford, Flamsteed had still not published his catalogue of the northern stars, nearly thirty years after Halley had published his (admittedly more limited) *Catalogue of the Southern Skies*. Some idea of what people felt about this can be gleaned from a letter written by James Gregory (David's brother) in 1699:

> Mr Flamsteed has rectified above 3,000 fixed stars; but is so perversely wicked that he will neither publish nor communicate his observations.

This was not a situation that Isaac Newton, the new President of the Royal Society, was willing to tolerate. He gained a formal promise from Prince George, the Consort of Queen Anne, that he would pay for the publication of the catalogue, which Flamsteed claimed was ready for the press, or would be once a fair copy was made. A Committee of Referees, headed by Newton (but not including Halley), was established to check the manuscripts and get them ready for publication, and Flamsteed handed over a rough copy of his observations on the understanding (as he saw it) that nothing would be printed until he provided a final fair copy. But that is not what the Referees understood, and they set the ball rolling, albeit somewhat sluggishly. David Gregory had the main responsibility for overseeing the project, and printing started in May 1706 but in a very desultory manner, and stopped entirely in 1708 when both Gregory and Prince George died. Nothing more happened until 1710, when Anne appointed a Board of Visitors (headed, of course, by Newton, and this time including Halley) to oversee the work of the Royal Observatory. As well as being empowered to demand from Flamsteed a 'true and fair copy' of his observations each year, the Visitors had the power to instruct him what observations to make, and to inspect his instruments to see that everything was being kept in good order. This was more than a mere slap on the wrist for Flamsteed, but the work they were particularly interested in was the catalogue; in December 1710 Halley was given the job of seeing the project through, adding his own observations and computed results to fill in gaps in the material supplied by Flamsteed. This was entirely within the remit of the Visitors, and Halley tried to keep Flamsteed in the loop, even sending him proofs for checking, with a conciliatory covering letter ending 'if you signify what's amiss, the errors shall be noted, or the sheet reprinted, if the case require it. Pray govern your passion, and when you have seen and considered what I have done for you, you may perhaps think I deserve at your hands a much better treatment than you for a long time have been pleased to bestow on Your quondam friend, and not yet profligate enemy (as you call me).' But Flamsteed still refused to cooperate.

While the catalogue was going through the press, in October 1711 Flamsteed had a furious row with Newton at a meeting called to discuss the need for new instruments at the Greenwich Observatory, and ended up calling Newton an atheist – which we know, as Flamsteed did not, would have been particularly barbed. That was the end of any prospect of friendly relations between Newton and Flamsteed. Halley fared no better. Flamsteed continued to refuse all overtures of friendship, and after Halley visited him in June 1712, taking with him companions clearly carefully chosen to reduce the prospect of unpleasantness, Flamsteed recorded in his notes:

the impudent editor, with wife son and daughters* attending him, and a neighbouring clergyman in his company, came hither, I said little to him. He offered to burn his catalogue (so he called the corrupted and spoiled copy of mine, of

---

* The son and daughters were now in their twenties.

which I had now a correct and enlarged edition in the press, and the second sheet printing off) if I would print mine. I am apt to think he knew it was so, and was endeavouring to prevent it. But to render his design ineffectual, I said little to him of it: so he went away not much wiser than he came.

The upshot of the unpleasantness was that the Gregory/Halley version of the catalogue appeared in 1712, but Flamsteed and his wife managed to buy up most of the copies and burn them. Flamsteed, now in his late sixties, was, however, at last moved to put his data in order for publication (the reference to it being already in the press in 1712 seems to have been an exaggeration), and it was eventually published in three volumes in 1725, giving three thousand star positions to an accuracy of 10 arc seconds, much better than earlier catalogues. By then, though, Flamsteed was dead, and Halley had succeeded him as Astronomer Royal.

Catalogues such as those of Flamsteed, Hevelius and Halley were intended as charts of the positions on the sky of the fixed stars, thought to be eternal and unchanging. But during his time as Savilian Professor, Halley realised that this is not always the case. 'New' stars, or novae, that had appeared from time to time, flaring up brightly then fading into obscurity, had already proved that the heavens were not entirely unchanging, but Halley seems to have been the first person to recognise that there are two kinds of nova. Writing in the *Philosophical Transactions* in 1715, he noted that the new star seen by Tycho Brake in 1574 and the one seen by Kepler in 1604 were particularly bright and faded very quickly, while other novae were less bright, and some of them seemed to flare up every few years. 'These two,' wrote Halley, 'seem to be of a distinct Species from the rest'; his was the first recognition of what we now call supernovae. A couple of years later, when he was already over sixty years old, Halley reported the profound discovery we mentioned. He found that even the 'permanent' stars do not all stay in the same place.

Halley was interested in the way the pattern of the stars seen from Earth seems to drift in the long term, which is a result of the way the Earth wobbles as it orbits the Sun. In the course of this work, while preparing astronomical tables, Halley compared the latest catalogues he had available with a star catalogue that had originally been compiled by Hipparchus in the second century Bc and had been recorded in the works of Ptolemy. Most of the star positions in the old catalogue matched those of the same stars in the catalogues made in Halley's day, but three stars were in significantly 'wrong' positions. Halley realised that the old observations were accurate – as shown by the positions of the other stars – but that a few stars had moved strikingly across the sky since the time of Hipparchus. The star Arcturus, for example, had moved across the sky by about twice the width of the full Moon in just under two thousand years, far too much to be explained as a mistake by the Greek astronomer. Halley pointed out (in the *Philosophical Transactions*) that the stars must be at different distances from us, and that the ones we see moving in this way must be the nearest ones; while more distant stars must be moving in the same way, we do not notice this because of the greater distances.

> What shall we say then? It is scarce credible that the Antients could be deceived in so plain a matter . . . these Stars being the most conspicuous in Heaven, are in all probability nearest to the Earth, and if they have any particular Motion of their own, it is most likely to be perceived in them, which in so long a time as 1800 Years may shew itself by the alteration of their places, though it be utterly imperceptible in the space of a single Century of Years.

The discovery of stellar motion was one of the most important observational discoveries in the history of astronomy, and a completely new idea in the eighteenth century, but seems to have passed almost without comment at the time. Halley's attention, though, was not solely fixed on the distant stars. Around the same time he was making this discovery, Halley was also demonstrating his impressive organisational skills and continuing intellectual power with observations of an event much closer to home, an eclipse of the Sun. These observations became part of a legacy that he would leave to future generations, in the same way that Hipparchus had provided a legacy of data for Halley.

Astronomers knew that there would be a solar eclipse visible from England on 22 April 1715, and Halley was determined that the best use should be made of this opportunity to observe such an event in detail. In 1714, he published a map showing the predicted path of the shadow of the Moon across England and Wales.* The map was partly to reassure a populace still superstitious about such things that the eclipse was a natural event and nothing to worry about. The chart carried a reassuring explanation:

> The like Eclipse not having not for many Ages been seen in the Southern Parts of Great Britain, I thought it not improper to give the Publick an Account thereof, that the suddain darkness wherein the Starrs will be visible about the Sun, may give no surprize to the People, who would, if unadvertized, be apt to look upon it as Ominous, and to Interpret it as portending evill to our Sovereign Lord King George . . .

But the main purpose of publishing the chart was scientific;

> . . . The Curious are desired to Observe it, and especially the Duration of Total Darkness, with all the care they can; for thereby the Situation and dimensions of the Shadow will be nicely determined; and by means thereof we may be enabled to Predict the like Appearances for ye future, to a greater degree of certainty than can be pretended to at present, for want of such Observations.

Halley wanted to record the exact track of the eclipse across the country, and in addition to this general plea he wrote to several

---

* Flamsteed, as Astronomer Royal, also published his prediction of the eclipse path, but this turned out to be less accurate than Halley's.

reliable people who were in the path of the eclipse, or better yet on the edge of the predicted path of the shadow, to ask them to make careful observations, which were then sent to the Royal and collated by Halley. This was the first systematic, large-scale observation of a solar eclipse.

Halley himself observed the event from Crane Court in London, in the company of several Fellows and their distinguished guests, including visitors from overseas. One of the guests was Isaac Newton's niece, Kitty Barton, who looked after Newton's household. She was a noted beauty, and made a deep impression on one of the French visitors, Rémond de Montmort, who later wrote from Paris thanking Newton for presents 'given by Mr Newton and chosen by Mrs Barton whose wit and taste are equal to her beauty'. She was, indeed, known in smart circles in London as 'pretty, witty, Kitty'. But Halley's eyes were on the heavens, not on her.

The observations were, for once, blessed by good weather. In 1715, Halley published his account of the observations of the eclipse made by his scattered reporters, giving particular emphasis to the observations from near the edge of totality. He calculated that the southern edge of totality passed between Newhaven and Brighton on the Sussex coast, and between Herne and Reculver on the north Kent coast, while the equivalent northern line ran from Haverfordwest to Flamborough Head. The key to this aspect of Halley's legacy was provided by the observations from places where the eclipse was just total, at the very edge of the shadow, as we shall discuss shortly.

A year later Halley made another contribution to his scientific legacy, when he published a paper discussing the possibility of determining the distance to the Sun by observing a transit of Venus. He had been interested in the idea for decades, and had read a paper on the subject to the Royal in 1691, but the 1716 paper in the *Philosophical Transactions* was his definitive last word on the subject. The idea itself was not new, but Halley's careful prescription of what should be done was. It included a detailed explanation of how and where the necessary observations should be made, and a plea that astronomers should take advantage of the opportunity to make such observations in 1761 and 1769, the dates of the next two such transits, by which time Halley himself would be dead. We shall describe what happened then in due course.

Halley's most famous legacy, though, concerns the comet that now bears his name. As we have discussed, he was calculating cometary orbits back in the 1690s and published most of this analysis in 1705; the final version appeared in 1726 but differs only slightly from the earlier version. Halley calculated that the same comet had been seen in 1531, 1607 and 1682, following an elongated elliptical orbit around the Sun, seventy-six years long, in accordance with the inverse square law of gravity. He predicted that the comet would be seen again round about the end of December 1758, and urged astronomers to watch out for it, 'whereof if according to what we have said it should return again about the year 1758, candid posterity will not refuse to acknowledge that this was first discovered by an Englishman.' As we shall see, his wish for due credit was granted, perhaps even more fully than he could have anticipated. But first we should describe Halley's later life as Astronomer Royal and the grand old man of British science.

By the time he became Astronomer Royal, Halley had been granted the only academic distinction that he ever seemed to care about. In 1710 he was awarded the degree of Doctor of Civil Laws by the University of Oxford. Up until that time, he had preferred the title 'Captain' to 'Professor', but from then on, he was happy to style himself 'Doctor' Edmond Halley. He was particularly pleased with the honour because it was made in recognition of his public service, not least his public service as a sea captain. The recommendation from the Chancellor of the University to the University Convocation read:

> Mr Edmund Hally having been Master of Arts near thirty years and often employed by her Majesty and her Predecessors in the Service at Sea in the remotest parts of the World to the

great Satisfaction of the Lords of ye Admiralty and others of the first Quality in the nation; And now being Your Professor of Geometry and well known to be a Person of great knowledge not only in that Science but in most of ye other parts of Learning, I have thought fitt in consideration of his great merit and the Service he hath done to the public both at home and abroad to recommend it to You that you would confer the Degree of Dr of Civil Laws on him without fees . . .

The mention of fees being waived is a significant indication of the esteem in which Halley was held; when Handel visited Oxford in 1733 to put on concerts, he declined the offer of an honorary degree because he was asked to pay a fee. The mention of services to 'others of the first Quality' also seems significant, hinting once again at services that are not a matter of public record. So it would be as Dr Halley that Edmond succeeded John Flamsteed as Astronomer Royal when Flamsteed died on 31 December 1719, somewhat to the relief of all those who had been frustrated by his general bloody-mindedness and unwillingness to make his observations widely available.

Flamsteed's death was not unexpected. He was seventy-three, and not in good health. Nor was Halley's appointment as the second Astronomer Royal unexpected. Indeed, it seems to have been anticipated, since on 11 December, nearly three weeks before Flamsteed died, Thomas Hearne wrote to Richard Mead, one of the Visitors, 'I must heartily congratulate you upon the success you have had on behalf of Dr Halley. This great Man had been neglected too long.' On 10 January Hearne recorded that 'Mr Flamsteed is dead and Dr Halley hath got his place at Greenwich'. Halley himself was now sixty-three, and moved quickly to install himself at the Observatory, ejecting Mrs Flamsteed, who took with her most of the instruments, claiming with some justification that they were Flamsteed's property. So the first task of the new Astronomer Royal was to re-equip the Observatory, which he did, with the aid of a Treasury grant of £500, with much better instruments than those used by his predecessor.

It might have been expected that Halley himself, in view of his advancing years, would make little use of the instruments. But on the contrary, in his sixty-fourth year he at last had an opportunity to begin a programme of observing the Moon over its entire eighteen-year Saros cycle. Maybe he was optimistic; maybe he expected his own successor to complete the task. But as it happens, he lived long enough to finish the job, his major observational achievement as Astronomer Royal. But he also lived long enough to see, and be involved in, the technology that would make this method of determining longitude obsolete.

In 1714, the British government had offered a prize of £10,000 for anyone who could devise a method to determine longitude at sea so accurately that it would produce an error of less than sixty nautical miles on a voyage to the West Indies and back, rising to £15,000 if the error was less than forty nautical miles and £20,000 if the error was less than thirty nautical miles. It was widely expected that an astronomical technique would eventually solve the problem. But at the end of the 1720s (the exact date is not certain), a young carpenter and self-taught clockmaker called John Harrison* visited Halley at Greenwich to show him a clock mechanism that he had devised, which he thought could be developed into a practical timepiece that would win the prize. Halley encouraged Harrison and introduced him to a leading clockmaker and Fellow of the Royal Society, George Graham. Graham not only gave encouragement but provided Harrison with £200, which Harrison and his brother James used to construct a wooden chronometer using his ideas (how Hooke would have loved it!). It was tested by the Royal Navy on ships travelling to Lisbon and back, producing spectacular results, and Halley was a member of the Royal Society board that officially endorsed Harrison's project. There followed a long saga of development, with Harrison repeatedly frustrated by the doubts of the authorities, until more than twenty years later he produced his final, successful design (and even longer before he received part of the prize, but not all

---

* No relation to Halley's former lieutenant.

of it). James Cook carried one of Harrison's chronometers on his second voyage, in 1772.

The rest of Halley's life can be summarised briefly. He had served as Secretary of the Royal Society since 1713, but resigned in 1721 (he did not have to resign his Oxford professorship on becoming Astronomer Royal, although this seems to have become something of a sinecure). But he maintained contact with the Royal, attending meetings and the informal gatherings in coffee houses and taverns after meetings, and serving as Vice-President in 1731. In 1729, Queen Caroline, the consort of George II (whom Halley had met on his visit to Hanover) visited the Royal Observatory, and was intrigued to learn that Halley had served as a Royal Navy captain. We don't know how much, if at all, Halley pressed his case, but she learned that as he had served more than three years in this capacity he was entitled to a pension of half-pay. Shortly after the visit, the King personally ensured that this pension was paid, and Captain Halley received it for the rest of his life.

In January 1736 Halley's wife, Mary, died (there are no details of the circumstances), and he seems to have suffered a slight stroke soon afterwards. This affected his right hand, but did not stop him observing, with the aid of an assistant. His son, another Edmond, died in February 1741, leaving no recorded children; Halley himself declined in strength through 1741, his eighty-fifth year, and died on 14 January 1742 (25 January New Style), quietly, sitting in a favourite chair after a glass of wine, and was survived by his two married daughters, Margaret and Katherine.

He was also survived by a body of work which still included unfinished business, a legacy for the next generation of astronomers and beyond. The first of these projects to bear fruit was the famous prediction of the return of 'his' comet.

As Halley and Newton had appreciated, predicting the exact time of the return of the comet depended on working out the perturbing influences of Jupiter and Saturn. The Frenchman Alexis-Claude Clairaut was the first person to devise a way to tackle the problem using a development of Newton's equations, but even with the equations the problem still had to be solved numerically. This was a horrendous task involving lengthy and tedious arithmetical calculations, but it was carried out by the French astronomer Jérôme Lalande,* with very considerable assistance from the splendidly named Mme Nicole-Reine Étable de la Brière Lepaute, a friend of Clairaut, who seems to have been an expert at astronomical calculations. Clairaut announced the result of these calculations – which took the pair more than six months to complete – to the French Academy of Sciences on the 14 November 1758, predicting that the closest approach of the comet to the Sun (perihelion) would occur in April 1759.

The comet was first seen (on this visit) by a German astronomer, Georg Palitzsch, on Christmas Day 1758, but news of his observations did not spread quickly, and it was 'discovered' independently by the French astronomer Charles Messier, working at the Paris Observatory, on 21 January 1759. Perihelion actually occurred on 13 March that year, so the official date for the return is now given as 1759, not out of any disrespect to Palitzsch, but because perihelion is an accurate and unambiguous date, whereas the exact timing of the first observation of a returning comet is, as this example shows, largely a matter of luck.

But whoever saw it first, the success of the prediction was sensational news. Lalande wrote:

The universe beholds this year the most satisfactory phenomenon ever presented to us by astronomers; an event which unique until this day changes our doubts to certainty and our hypotheses to demonstration.

It was the fact that this was a successful *prediction* that made the observations so important. It was all very well Newton (with a little help from Hooke) explaining why the orbits of the planets were elliptical, but astronomers already knew that the orbits were elliptical. And even before that, it was known that the planets

---

* In full, Joseph Jérôme Lefrançois de Lalande.

were regular and predictable in their movements. But comets were the archetypal unpredictable phenomenon, appearing entirely without warning, rousing superstitious awe in the eighteenth century to an even greater extent than eclipses. Accurately predicting the date on which a comet would appear in the sky was the greatest triumph of the scientific method up to that time, and it established the 'Newtonian' way of understanding how the world works. The predicted return of the comet confirmed the accuracy of the inverse square law (suspected by others and proved by Newton), that gravity is an attractive (centripetal) force as Hooke had pointed out to Newton, and that the laws that govern the Universe are, as Hooke had also pointed out, the same as the laws which apply here on Earth. This world-view, which 'changes our doubts to certainty and our hypotheses to demonstration', owed at least as much to Hooke as to Newton, and its successful application was down to Halley. All three of them were dead by 1759, but their scientific legacies survived, and the scientific truth lived on. Where next could this understanding of the laws that govern the Universe take astronomy? Halley had already pointed the way there, as well.

The first astronomer to predict and observe a transit of Venus was a young Englishman called Jeremiah Horrocks. He was born in 1618, near Liverpool, and attended Emmanuel College, Cambridge, from 1632 to 1635, but like many of his contemporaries did not bother graduating. Horrocks was a dedicated amateur astronomer (there were no professional astronomers in England then) who also worked in the family business of watch- and instrument-making. In 1629, Johannes Kepler had published a pamphlet in which he predicted, based on the astronomical tables he had produced, that there would be a transit of Mercury in 1631, and transits of Venus in 1631 and 1761, with a 'near miss' in 1639. Horrocks was too young to observe the transits in 1631, and nobody else seems to have bothered, but when he reworked Kepler's calculations using the best available data he found that Kepler had made a tiny error and there would, in fact, be a transit of Venus in 1639, which he predicted would occur at about 3 pm on 24 November on the Julian calendar (4 December on the Gregorian calendar). He was able to observe the event, using the standard technique of projecting an image of the Sun through a telescope on to a white surface, and his friend and colleague William Crabtree made similar observations. Horrocks wrote an account of the event in Latin (*Venus in sole visa*), but his death from unknown causes early in 1641 at the age of twenty-two meant that it was not published in his lifetime. However, Johannes Hevelius was sent a copy, and published it in 1661, causing great interest at the Royal Society, among other places.

Following Horrocks' work, it was clear that transits of Venus always occur in pairs, following a pattern that repeats every 243 years. The two transits in a pair are separated by eight years, then there is a gap of 121.5 years followed by another pair of transits and a gap of 105.5 years before the whole pattern repeats. So the total length for the cycle is $(8 + 121.5 + 8 + 105.5) = 243$ years.

In 1663 James Gregory, the Scottish mathematician who invented the Gregorian telescope design, pointed out that by observing and carefully timing a transit of Mercury from two widely separated points on the surface of the Earth it ought to be possible to calculate the distance to the Sun using geometry, essentially the same system of triangulation, involving parallax, that is used by surveyors to calculate distances to objects without ever visiting them. The essential point is that the widely separated observers see the planet pass in front of the Sun from slightly different angles, so that they record slightly different times for key moments of the transit, such as the time when the edge of the planet first seems to touch the edge of the disc of the Sun and the duration of the transit. With this information and the precise locations of the observing sites it is possible to calculate the angles involved (too small to measure directly) and work out the distances. Although Halley observed a transit of Mercury from St Helena in 1677, there was no concerted effort by astronomers to make observations from other locations, and there was insufficient data to put the idea into practice. But it sowed the seed for Halley's publications of 1691 and 1716 urging for a proper investigation of the pair of transits of Venus due in

1761 and 1769, and providing a detailed plan both for the necessary observations and the technique for using the results to work out the distance to the Sun. Venus is a better choice for the job, because it approaches much closer to the Earth than does Mercury (or, indeed, any other planet), although before the job was carried out nobody knew exactly how close. But even in the case of Venus, a transit lasting typically for around six hours has to be timed to an accuracy of about 10 seconds for the parallax effect to be noticeable.

By 1761, Europe was again at war. This one was known as the Seven Years' War, although it actually lasted from 1754 to 1763; however, there had already been a Nine Years' War, and this time most of the fighting was confined to the period from 1756 to 1763. Nevertheless, this did not stop several countries making plans to observe the transit – if anything, the war spurred competition among the belligerents (notably Britain and France) to outdo each other. Indeed, the British effort was a badly funded project hastily cobbled together when the Royal Society learned that the French were planning something. Halley had suggested that observations should be made from sites such as Hudson's Bay, Norway and the Far East. This broad spread of observing sites was possible because the event occurred in northern summer, when the nights are short. This meant that observers at high northern latitudes could watch the beginning of the transit before sunset, and then pick up the end of the transit after the short northern night, while observers on the far side of the world simultaneously made observations close to midday. But the plans were partially frustrated by the war.

The British expedition to make observations on the far side of the world, at Sumatra, set off on board HMS *Seahorse* on 8 January 1761, while a second team was sent to St Helena. The two observers on board the *Seahorse* were Jeremiah Dixon and Charles Mason, later to become famous as surveyors of the Mason–Dixon line in what became the United States. But only a couple of days after leaving port the ship was attacked by a French frigate and suffered severe damage, limping back to harbour with eleven dead and thirty-seven wounded. After repairs they set off again, but because of the delay had to make their observations from the Cape of Good Hope. Altogether, well over a hundred observers were involved in observations of the transit of 1761, but without measurements from the Pacific the data were not of the best quality. This made astronomers all the more determined to take full advantage of the 1769 transit, with national pride seen to be very much at stake. In 1766, promoting the possibility of an expedition to the Pacific, the Royal boasted that the British 'were inferior to no nation on earth, ancient or modern'.*

Just over a year later, when plans for the expedition were well advanced, an unexpected bonus fell into their lap. The Cornish navigator Samuel Wallis returned to England with his ship the *Dolphin*, having circumnavigated the globe, and announced his discovery of the island now known as Tahiti. It was already vaguely known to mariners that there were Pacific islands, but Wallis was the one who stopped at Tahiti and made an accurate determination of its latitude and longitude. His return was just in time for Lieutenant James Cook, Captain of HMS *Endeavour*, to get the news (and some of Wallis's crew members) before setting off on his own circumnavigation. Tahiti was clearly the best place known for the transit observations he intended to make, and his plans were revised accordingly.†

Cook was not only Captain of the *Endeavour*, but an experienced and competent astronomer who was one of the two officially nominated observers, the other being Charles Green; he even received a small stipend for his astronomical contribution on top of his naval pay. Green, who worked at the Royal Greenwich Observatory, was an expert at determining longitude using the lunar method, a key to determining the precise location of Tahiti. Cook's ship reached Tahiti on 13 April, and the crew set about building a substantial observatory within an enclosure known as Fort Venus on a sandy spit of land they dubbed Point Venus (still known by that name).

---

* See R. and T. Rienits, *The Voyages of Captain Cook*, Hamlyn, London, 1976.

† Hawaii, which would have been an even better location, only became known to Europeans when Cook himself found it on his second voyage.

Some idea of the scale of the fortified enclosure can be gleaned from the fact that fifty-four tents were contained within its walls, housing not only the men but a blacksmith's workshop and cookhouse as well as the observatory itself. As the day of the transit, 3 June 1769, approached, parties were also dispatched to the other side of the island and to a nearby island, Moorea, for additional observations.* Although, as Cook noted in his journal, the day 'prov'd as favourable to our purpose as we could wish, not a Clowd was to be seen the whole day and the Air was perfectly clear, so that we had every advantage we could desire in Observing the whole of the passage of the Planet Venus over the Suns disk', the different observers found that their measurements of the exact moments when Venus entered and left the Sun's disc 'Differ'd from one another . . . much more than could be expected.'

The Royal was disappointed by the results, and placed the blame for what they saw as an at least partial failure of the expedition on Green, who conveniently had died on the way home and could not answer back. But in 1771 the Savilian Professor of Astronomy, Thomas Hornsby, published an analysis of all the 1769 transit observations, which led him to conclude that 'the mean distance from the Earth to the Sun [is] 93,726,900 English miles', and in the same year Jérôme Lalande, who was also involved in the prediction of the return of Halley's comet, used data from both the 1761 and 1769 transits (from sixty-two observing stations in 1761 and sixty-three, mostly different, observing stations in 1769) to come up with a distance of 153 million kilometres, plus or minus 1 million km. The modern value is 149,597,870 km (in round, imperial, numbers, 92,955,000 miles), and it seems likely that Hornsby got such a 'good' value as much by luck as by skill, with some errors cancelling each other out. But even Lalande's estimate is impressive for the time. With the aid of Kepler's laws, it meant that the distances of all the planets from the Sun could be calculated with reasonable accuracy for the first time.

---

* To put the date in another context, this was a few weeks before the birth of Napoleon Bonaparte, on 15 August 1769.

The most important aspect of the observations of the two transits of Venus in the 1760s was not, however, the measurement of the distance to the Sun. It was the fact that the effort was made at all. This was the first time there was an organised international effort to tackle a major scientific problem, at great expense. It has been likened in scope with the European collaboration on the Large Hadron Collider at CERN in modern times. But the scientists at CERN do not have to battle with enemy frigates, storms at sea, scurvy and tropical diseases. They can reasonably expect that, unlike Green, they will get home after their experiments.

Cook may have been disappointed by the limited success of the transit observations, but his voyage was far from over. At Tahiti, he opened sealed orders from the Admiralty, which revealed why the Navy and the Government had been willing to subsidise this scientific expedition, contributing not only the ship but also £4,000 towards the costs. He was to head south in search of the fabled Terra Australis Incognita (unknown land of the south), and if he failed to find it to turn west and investigate New Zealand and New Holland (Australia), recently discovered by the Dutch. The rest, as they say, is history. Perhaps those territories would have fallen into British hands anyway, but it was the timing of the transit, and Halley's promotion of the idea of an expedition to the Pacific region to make observations, that made it happen when it did.

Accounts of Halley's work usually end here, saying that this was his last contribution to both science and discovery, made two decades after his death. But there is more. One of Halley's 'experiments' actually did not achieve fruition until the end of the 1970s, not two decades but two *centuries* after his death, and once again it involved the Sun.

In the 1970s, the American astronomer Jack Eddy, working at the High Altitude Observatory of the US National Center for Atmospheric Research, became interested in the possibility of a link between climate change on Earth and changes in the level of solar activity, measured by the number of dark spots seen on the surface of the Sun each year. Over a cycle roughly eleven years long, the number of sunspots first increases to a peak then decreases to a minimum, but the peak level reached differs from one cycle to another. It happens that during the second half of the seventeenth century, exactly at the time of severe European cold mentioned earlier (Chapter Six), the Sun had been very quiet, with hardly any sunspots seen between 1645 and 1715. This interval of low solar activity became known as the Maunder Minimum, after the astronomer who first noticed it from his studies of old records. This led Eddy and his colleagues to investigate the possibility that the size of the Sun mighty be changing – that it was either shrinking or expanding, very slowly – and that this was affecting its output of heat. One way to test this was by looking at old records of transits of Mercury, to see how long it took for the planet to cross the disc of the Sun. If the Sun was larger in the past, for example, then the transit would have taken longer. But the observations were not precise enough to settle the question. There was, though, one set of very precise old data that could be used to test the idea: Halley's study of the eclipse of 1715.

The exact position of the edge of the shadow on the Earth cast by the Moon during a solar eclipse depends on the size of the Moon (which is well known and doesn't change) and on the exact position of the Moon in its orbit during the eclipse, which determines its distance from us and can be calculated. It also depends on the exact size of the Sun. At the end of 1980, a team of researchers headed by David Dunham reported a comparison of Halley's records of the 1715 eclipse with data from an eclipse seen in 1976 in Australia and one visible from North America in 1979. They concluded that 'between 1715 and 1979, a decrease in the solar radius of 0.34 + 0.2 arc seconds was observed'. This stimulated further investigations using a variety of historical and modern data, including transits and other eclipse observation, which suggested that there has been a long, slow decline in the radius of the Sun of 0.01 per cent (about 44 miles/70 km) since the early 1700s. If this is a real phenomenon (and it is still possible that such a small apparent change might simply be appearing because some of the old records are inaccurate) it is probably part of a long-term cycle in which the Sun 'breathes' in and out; it does not mean that the Sun is going to shrink away and disappear. Whether or not the change is real, the links between changes in solar activity and climate are still unproven and a matter of (sometimes heated) debate among the experts. But that is not our concern here. What matters in the present context is that not only was Halley far-sighted enough to provide valuable data for future generations to use, he was good enough at his job to make sure that those data really were useful, 238 years after his death. Perhaps Dunham and his colleagues should have included Halley as co-author of their paper. But that really does mark the end of the direct contributions of Halley and Hooke to science.

How can we sum up the relative achievements of Hooke, Halley and Newton, and their contribution to the scientific revolution? Ironically, in view of Newton's religious beliefs, the best approach is to treat them as a trinity. Hooke had the greatest physical insight, and even if we set to one side his other scientific achievements (microscopy, geophysics and the rest he was the first person to realise that the same laws of physics apply in the Universe at large as here on Earth, and to appreciate in particular that the inverse square law of gravity is a universal force and that it acts centripetally; Newton was a mathematical genius (his other activities, alchemy and theology, are best set to one side) who codified the new physics by providing a set of equations to describe the behaviour of everything from balls rolling down slopes to planets orbiting the Sun; Halley (apart from his other achievements as an astronomical observer and geophysicist) was the first person to apply those equations to new problems, rather than 'merely' explaining past observations, and use them to make successful predictions, the ultimate (indeed, only) test of any scientific theory. None of them deserves to be remembered in the shadow of any of the others, but if push came to shove, we would certainly place Hooke 'first among equals'.

CODA

HOW TO DO SCIENCE

In an undated document, published after his death* but probably drawn up early in his career as Curator of Experiments, Robert Hooke set out his prescription for an ideal way to carry out scientific investigations. This not only gives insight into the way he, and the Royal Society, worked, but is relevant to anyone planning to do science today, or to understand why science provides the best description we have of the way the world works.

The Reason of making Experiments is, for the Discovery of the Method of Nature, in its Progress and Operations.

Whosoever therefore doth rightly make Experiments, doth design to enquire into some of these Operations; and, in order thereunto, doth consider what Circumstances and Effects, in that Experiment, will be material and instructive in that Enquiry, whether for the confirming or destroying of any preconceived Notion, or for the Limitation and Bounding

---

* See Derham/Hooke, 1726.

thereof, either to this or that Part of the Hypothesis, by allowing a greater Latitude and Extent to one part, and by diminishing or restraining another Part within narrower Bounds than were at first imagin'd, or hypothetically supposed.

The method therefore of making experiments by the Royal Society, I conceive, should be this.

First, to propound the Design and Aim of the Curator in his present Enquiry.

Secondly, to make the Experiment, or Experiments, leisurely and with Care and Exactness. Thirdly, to be diligent, accurate, and curious, in taking Notice of, and shewing to the Assembly of Spectators, such Circumstances and Effects therein occurring, as are material, or at least, as he conceives such, in order to his Theory.

Fourthly, After finishing the experiment, to discourse, argue, defend, and further explain, such Circumstances and Effects in the preceding Experiments, as may seem dubious or difficult: And to propound what new Difficulties and Queries do occur, that require other Trials and Experiments to be made, in order to their clearing and answering: And farther, to raise such Axioms and Propositions, as are thereby plainly demonstrated and proved.

Fifthly, To register the whole Process of the Proposal, Design, Experiment, Success, or Failure; the Objections and Objectors, the Explanation and Explainers, the Proposals and Propounders of new and farther Trials; the Theories and Axioms, and their Authors; and in a Word, the History of every Thing and Person, that is material and circumstantial in the whole Entertainment of the said Society: which shall be prepared and made ready, fairly written in a bound Book, to be read at the Beginning of the Sitting of the said Society: the next Day of their Meeting, then to be read over, and further discoursed, augmented or diminished, as the Matter shall require, and then to be sign'd by a certain Number of the Persons present, who have been present, and Witnesses of all the said Proceedings, who, by Subscribing their Names, will prove undoubted testimony to Posterity of the whole History.

Which, of course, is the practical application of the Society's motto, *Nullius in Verba*.

BIBLIOGRAPHY

Angus Armitage, *Edmond Halley*, Nelson, London, 1966.

John Aubrey, *Brief Lives* (edited by Andrew Clark), Clarendon Press, Oxford, two volumes, 1898.

Jim Bennett, Michael Cooper, Michael Hunter & Lisa Jardine, *London's Leonardo*, Oxford UP, 2003.

Thomas Birch, *The History of the Royal Society of London*, four volumes published 1756–57.

Savile Bradbury, *The Evolution of the Microscope*, Pergamon, Oxford, 1967.

F. F. Centore, *Robert Hooke's Contribution to Mechanics*, Martinus Nashoft, The Hague, 1970.

Alan Cook, *Edmond Halley*, Oxford UP, 1998.

Michael Cooper, *Robert Hooke and the Rebuilding of London*, Sutton, Stroud, 2003.

J. G. Crowther, *Founders of British Science*, Cresset Press, London, 1960.

Clara de Milt, 'Robert Hooke, Chemist', *Journal of Chemical Education*, volume 16, pp 503–519, 1939.

William Derham, *Philosophical Experiments and Observations of the Late Eminent Dr. Robert Hooke*, originally published 1726, Cassell, London, 1967. See also Hooke, below.

Kerry Downes, *Christopher Wren*, Allen Lane, 1971.

Ellen Tan Drake, *Restless Genius: Robert Hooke and his Earthly Thoughts*, Oxford UP, 1996.

Margaret 'Espinasse, *Robert Hooke*, Heinemann, London, 1956.

John Evelyn, *Diary*, edited by E. S. de Beer, Clarendon Press, Oxford, 1955.

E. G. Forbes, A. J. Meadows & D. Howse, *Greenwich Observatory*, 3 volumes, Taylor & Francis, London, 1975.

John Gribbin, *Science: A History*, Allen Lane, London, 2002.

R. T. Gunther, *Early Science in Oxford*, volume 6 & volume 7, 1930, volume 8, 1931, volume 10, 1935.

A. Rupert Hall, *Isaac Newton*, Blackwell, Oxford, 1992.

Robert Hooke, *Micrographia*, Royal Society, London, 1665. Facsimile edition, Dover, New York, 1961.

Robert Hooke, *Lectures and Discourses on Earthquakes*, reprinted from the Posthumous Works (edited by Waller) in an edition published by Arno Press, New York, 1978; also see the collected lectures in Drake.

Robert Hooke (edited by William Derham), *Philosophical Experiments and Observations of the Late Eminent Dr. Robert Hooke*, originally published 1726, reprinted Kessinger Publishing, Whitefish, Montana, 2010.

Robert Hooke, *Diaries*, see Robinson & Adams.

Michael Hunter, *The Royal Society and its Fellows*, British Society for the History of Science, Chalfont St Giles, 1982; revised edition 1994.

Michael Hunter & Simon Schaffer (editors), *Robert Hooke: New Studies*, Boydell Press, Woodbridge, 1989.

Stephen Inwood, *The Man Who Knew Too Much*, Macmillan, London, 2002.

Lisa Jardine, *The Curious Life of Robert Hooke*, HarperCollins, London, 2003.

Paul Kent & Allan Chapman (editors), *Robert Hooke and the English Renaissance*, Gracewing, Leominster, 2005.

Geoffrey Keynes, *A Bibliography of Dr. Robert Hooke*, Clarendon Press, Oxford, 1966.

Joseph Jérôme Lalande, *Tables astronomique de M Halley pour les planetes et les cometes, including 'l'histoire de la comete de 1757'*, Durand, Paris, 1759.

Rachel Laudan, *From Mineralogy to Geology: The Foundations of a Science, 1650–1830*, University of Chicago Press, 1987.

Charles Lyell, *Principles of Geology*, John Murray, London, 1830.

Eugene MacPike (editor), *Correspondence and Papers of Edmond Halley*, Clarendon Press, Oxford, 1932. Includes a biographical memoir of Halley, probably written by his contemporary Martin Folkes, who was President of the Royal Society from 1741 to 1752.

Eugene MacPike, *Dr Edmond Halley (1656–1742), A bibliographical guide to his life and work, arranged chronologically*, Taylor & Francis, London, 1939.

Isaac Newton, *Principia*, Royal Society, 1687; third edition in English (translated by Andrew Motte as *Mathematical Principles of Natural Philosophy and System of the World*) published in 1729 and available in a Cambridge UP edition published in 1934.

Richard Nichols, *The Diaries of Robert Hooke*, Book Guild, Lewes, 1994.

Samuel Pepys, *Diary* (edited by R. C. Latham & W. Matthews), eleven volumes published 1970–83 by Bell & Hyman, London.

Roger Pilkington, *Robert Boyle*, Murray, London, 1959.

Roy Porter, *The Making of Geology: Earth Science in Britain 1660–1815*, Cambridge UP, 1977.

Henry Robinson and Walter Adams, editors, *The Diary of Robert Hooke*, Taylor & Francis, London, 1935.

John Robison, in *On the Elements of Chemistry* (based on lectures given by Joseph Black), William Creech, Edinburgh, 1803.

Colin Ronan, *Edmond Halley*, Macdonald, London, 1970.

Thomas Sprat, *The History of the Royal Society of London*, Royal Society, 1665; facsimile edition published by Routledge, London, 1959.

William Stukeley, *Memoirs of Sir Isaac Newton's life*, CreateSpace Independent Publishing Platform, Kindle, 2016.

Norman Thrower, *The Three Voyages of Edmond Halley*, Hakluyt Society, London, 1981.

H. W. Turnbull (editor), *The Correspondence of Isaac Newton*, seven volumes, Cambridge UP, 1959–60.

Richard Waller (editor), *The Posthumous Works of Robert Hooke* (1705; facsimile available online at Google books).

Richard Westfall, *Never at Rest*, Cambridge UP, 1980.

See also:
http://www.newtonproject.sussex.ac.uk

INDEX